CAMPANULAS

Christopher Helm
HARDY PLANT series:

ORNAMENTAL GRASSES *Roger Grounds*

MECONOPSIS *James L. S. Cobb*

Forthcoming:

HARDY EUPHORBIAS *Roger Turner*

ALLIUMS *Dilys Davies*

DIANTHUS *Richard Bird*

The Hardy Plant Society

The Hardy Plant Society was formed to foster an interest in hardy herbaceous plants. It gives its members information about the wealth of both well known and little known hardy plants, how to grow them well and where they may be obtained.

Members receive regular bulletins and newsletters and there are regional and local groups, genus and special interest groups, a seed exchange, plant sales and shows. The Society has a special interest in plant conservation and has begun a programme of re-introducing desirable plants from the past.

More information may be obtained from the General Secretary, Hardy Plant Society, Garden Cottage, 214 Ruxley Lane, West Ewell, Surrey, KT17 9EU.

CAMPANULAS

Peter Lewis &
Margaret Lynch

Illustrated by
ROGER PHILLIPPO

Published in association with the Hardy Plant Society

CHRISTOPHER HELM
London

TIMBER PRESS
Portland, Oregon

Line illustrations by Roger Phillippo

Christopher Helm (Publishers) Ltd, Imperial House,
21–25 North Street, Bromley, Kent BR1 1SD

ISBN 0-7470-2009-4

A CIP catalogue record for this book
is available from the British Library

First published in North America in 1989 by
Timber Press, Inc.
9999 S.W. Wilshire, Portland,
Oregon 97225

ISBN 0-88192-150-5

Typeset by Paston Press, Loddon, Norfolk

Printed and bound in Great Britain by Biddles Ltd, Guildford and Kings Lynn

Contents

Colour Plates

Figures

Introduction

Campanulas have an appeal to most gardeners; it might be said that, however selective the gardener, there is a campanula to suit. The cottage gardener cannot do without them, the rock gardener looks on them as belonging to one of the few essential backbone genera; and the plantsman who loves a challenge for alpine house or show bench can find material enough to keep him busy. Whether the gardener grows for cutting, for flower arranging, for well furnished herbaceous bed or border, for alpine gardening, or even for bedding out, there is a bellflower which will amply suit and richly repay any time and effort applied.

There have been a number of accounts of the genus given over the years; these have been largely either in the botanical literature or in periodicals, and now when dug out of some archive, they have a very outdated look to them. Since Crook's monograph some 35 years ago there has been no overall treatment of the family at all, and in that time there have been a number of new discoveries.

In a genus of over 300 species, to which must be added more than a few variants and hybrids, there is ample scope for a substantial tome in which to assemble the available information, and further scientific research would be valuable. Until this is done, the present volume aims to describe the bellflowers that are available in cultivation, and to do this for the gardener and not the botanist. The Hardy Plant Society, of which we are Members, asked for a book to cover the hardy border flowers which it exists to promote. Thus border campanulas are the mainstay of this book; alpines are described, but in less detail.

We have also attempted to give some anecdotes of the history and discovery of various campanulas, as well as ideas on plant associations and on cultivation. In particular we have tried to be even handed and call a spade a spade—not all *Campanulas* are good tempered or easy—although most of them are.

We would scarcely either have got going, or kept going, but for the generous help of the University of Cambridge Botanic Garden, for access to their Cory Library. We also want to thank Clive King, Peter Orris, Peter Yeo and the administrative staff who have smiled rather than groaned when they have seen us looming up yet once more in quest of information.

Friends and experts have given us advice on the manuscript. In particular we should like to thank Will Ingwersen for giving us the benefit of his immense knowledge of the genus, and Mary Newnes for her comments about flower arranging.

A Brief History

In *Cymbeline*, Shakespeare extols 'The azure harebell, like her veins . . .'. But he was talking of what in England is called the Bluebell, *Scilla non-scripta*, an early spring bulb, and not a campanula. In Scotland, of course, the name bluebell is given to a campanula, *Campanula rotundifolia*, which in England is called a harebell. A great poet is at liberty to confuse a common name. The confusion of common names has continued to this day—and, in fact, has increased, when we add the bluebells of Australia and New Zealand, which although they are *Wahlenbergias* are close relatives.

Gerard

Gerard, in his herbal of 1597, was one of the first to tackle the name of *Campanulas*, or Bellflowers. He describes *Campanula medium* (now commonly called Canterbury Bells) thus: 'Viola mariana, Coventry Bells' which 'grow in woods, mountains and dark valleys and under hedges, especially about Coventry, where they grow plentifully abroad in the fields and are called Coventry Bells, and of some people about London, Canterbury Bells; but improperly for that there is another kind of bell flower growing in Kent about Canterbury which may more fitly be called Canterbury Bells, because they grow there more plentifully than in any other countrey.'[1]

Oh dear. If it were not for his excellent description we would not be at all sure what plant Gerard was on about. But he describes *Campanula medium* very clearly. 'Coventry Bells have broad leaves, rough and hairy . . . among which do rise up stiffe hairy stalks the 2nd year after the sowing of the seed, which stalks divide themselves into sundry branches, whereupon grow many fair pleasant bellflowers . . . cut to the brim with five slight gashes . . . in the middle of the flowers be 3 or 4 whitish chives and also much downie hair such as in the eares of a dog or such like beast . . .'. So Gerard's Coventry Bell is our Canterbury Bell.[1]

Gerard then goes on to describe Throatwort or Canterbury Bells. But here the Latin name gives us a clue, for he calls it *Trachelium majus*. It is our *Campanula trachelium*, a plant which we now in England call Bats-in-the-Belfry, and which he called Canterbury Bells. This is, by the way, one of the few campanulas which the old herbalists thought had any medicinal use. Gerard says 'We have found from our own experience that they are excellent good against the inflammation of the mouth and throat, to be gargled and washed with the decoction of them.'[1]

John Parkinson

John Parkinson, *Paradisi in Sole* 1629, was also keen to check out 'Coventry'. 'The Coventry Bells doe not grow wilde in any of the parts about Coventry, as I am credibly informed by a faithful Apothecary dwelling there, called Master Brian Ball, but are noursed in Gardens with them as they are in other places.'[2] As *C. medium* is not a British native this is clearly correct.

Parkinson goes on to describe '*Campanula persicifolia*, the Peach leaved Bell-flowers white or blew', whose name has thankfully not changed from then till now. He gave a useful tip: 'The roote is very small, white and thready, creeping under the upper crust of the ground so that oftentimes the heat and drought of the Summer will goe near to parch and wither it utterly; it requireth therefore to be planted in some shadowie place.'[2]

Good descriptions follow, of *Campanula pyramidalis* which he calls 'The Great or Steeple Bellflower'; *C. medium*, which he calls Coventry Bells; and *C. trachelium*, which he calls Canterbury Bells. He illustrates and describes the double forms of *C. trachelium* which are so much sought after today. Plants which are probably *C. latifolia* and *C. glomerata* are given good descriptions, as is the slightly related *Lobelia cardinalis*.

The virtues of *Campanulas* were not great, according to Parkinson. 'They may safely be used in gargles and lotions for the mouth, throate and other parts, as occasion serveth. The roots of many of them, while they are young, are often eaten in sallets by divers beyound the seas.'[2]

Carl Linnaeus

Carolus (Carl) Linnaeus took hold of *Campanula* and set it in proper order, when he published his *Genera Plantarum* in 1737. He confirmed the name, which has been followed ever since, as indeed has the rest of that account which was the start of modern systematic botany. He died, aged 70 in 1778, and at his funeral, 17 doctors of medicine, all his pupils, were pall bearers.

The Caucasus and Greece

If campanulas have an epicentre it must be in the Caucasus, with Greece running Eastern Europe a close second. Distribution is discussed in the next chapter, but it is interesting in the light of the lives of two very different men. Dr John Sibthorp, who gave his life to studying Greek flowers, and Marschall von Bieberstein, who lived in Russia and made his life's work Caucasian flowers. Sibthorp was born ten years earlier, but died when he was 38, while Bieberstein lived to be 58.

John Sibthorp

John Sibthorp was 25 in 1783 when he succeeded his father as Professor of Botany at Oxford. He was a very affluent young man, and arranged for a deputy to take over his job when three years later he went on his first botanical expedition. He took with him Ferdinand Bauer, a draughtsman, and travelled through the Mediterranean to Crete, then to Athens and Mount Olympus, and spent the winter in Constantinople. He visited 'the snowy heights of Parnassus, the steep precipices of Delphis, the empurpled mountain of Hymettus, the Pentele, the lower hills about the Piraeus, the olive grounds about Athens, and the fertile plains of Boetia.'[3] He and Bauer sailed home from Patras in September with 2,000 specimens, about 300 of them new, and a large number of drawings.[4] Sibthorp's health had not stood up to the heat in the islands and the rough journey home; but on this occasion he quickly recovered. Although he wished to return, pressure of work and politics kept him at home for seven years.

In 1794 he made another collecting trip. This started badly with a slow and stormy sea passage to Constantinople, where he arrived with a 'bilious fever and colic.'[4] But he went to Crete, climbed Mount Olympus, travelled to Troy and Mount Athos and back to Athens. He spent the winter on the Ionian island of Zante and from there, the next spring, visited the Peloponnese and climbed in the Taygete mountains. His journeys were fraught with hazards. He lost a young colleague, Francisco Borone, who fell from a window in Athens while sleep-walking; had to dodge barbary pirates; and then on the way home from Zante had another terrible sea voyage when bad weather prevented them from sailing for Italy. He arrived home with a fever and cough in the autumn of 1795, and died in Bath in February 1796.

Sibthorp is credited in *Hortus Kewensis* 1810 with 14 introductions, including *Campanula versicolor*, a plant from the area of Mount Olympus. Rather like a small *Campanula pyramidalis*, its thick woody stem hangs from the rocks. *Campanula versicolor* took a long time to reach the gardening public for, in 1913, Farrer was saying 'still out of reach'.[5] The dramatic flower is bicoloured light and dark blue, with a long projecting style.

Sibthorp left £30,000 to pay for the publication, posthumously, of his superb 10-volume *Flora Graeca*, with Bauer's plates, in 1806. Sir J. E. Smith, the editor, had no easy task as Sibthorp's notes were often scanty: 'He trusted to his memory and dreamed not of dying.'[6]

Bieberstein

Friedrich August Marschall von Bieberstein, born in 1768, came to Russia from a military academy in Stuttgart. He joined the Russian army and served in the Crimea for three years. When he left in 1796, he joined an expedition of Count Zubov's to Persia, from where he explored the Western Caucasus. Two years later he explored the north and east of the range.[4]

The first botanical work he wrote about this area, in French and German, was *Tableau des Provinces . . . entre les Fleuves . . . .* But much more important was *Flora Taurico-Caucasica*, written in Latin in 1808, which covered 2,322 species of which 17 were campanulas. These include *C. lactiflora*, which he introduced. He describes its favourite habitat as the Caucasian Alps, especially around the castle of Wladi-Kawkas on Mount Kaitchaur, where it flowered in September.[7] This is sufficient monument for any botanist. It has all the virtues: tall, elegant, and dripping with flowers. Bieberstein's *Flora* also included *Campanula adami* (now *C. tridentata*), which he may have named after a Dr J. M. F. Adams who, with Bieberstein, was a botanical member of Count Massin-Puschkin's journey in the Caucasus. The expedition was quite unofficial, but the members sent seed and specimens, which were important introductions, back to Sir Joseph Banks at Kew and to others. *Puschkinia scilloides* commemorates its leader.

Bieberstein went on no more organised collecting trips. But his new job, as inspector of silk-worm breeding,[4] took him all over Southern Russia from his home in Kharkov, and he had ample opportunity to look for Caucasian plants.

Fischer

Friedrich Ernst Ludwig von Fischer, 1782–1854, received a doctorate in his native Germany. But he made his name as the Director of the Imperial Botanic Garden at St Petersburg,[8] laid out by Peter the Great in 1714. As its new Director Dr Fischer travelled to England, France and Germany in 1824 to purchase exotic plants, and he returned with 2,320 species for the Garden. Dr Anderson at Chelsea Physic Garden received from him seed of *Campanula*

1 Dr F. E. L. von Fischer 1782–1854

latifolia 'Macrantha' in 1825. He also sent to Mr Hunneman in England seed of *Campanula sarmatica* and *Campanula speciosa*, which found its way to Kew.[8]

Dr Fischer gathered an enormous private herbarium of 60,000 species. His most important work was a catalogue of the plants in the Imperial Garden which listed many new species. In it, for instance, he mentions *Campanula grandis* (now *C. latiloba* or *C. persicifolia* subspecies *sessiliflora*), a fine garden plant. He also mentions *C. glomerata* var. *dahurica*, which is one of the best *C. glomerata* varieties. But sadly we have no description or note of where it was collected, although Dahurica is in the area of Lake Baykal in Siberia. Many of his herbarium specimens were inadequately labelled and did not give exact locations of the plants collected, saying, for instance, just 'China'[9]

In 1850 he was compelled to leave his post owing to irregularities in the accounts, and he died four years later. His herbarium was bought from his widow by Imperial order for 1,000 roubles.[9]

More Russians

It is possible that there are exciting campanulas still flowering unseen in the Caucasus: perhaps the Russians will find them. A novelty of just that rare and intractable sort that gardeners love was discovered in 1894—*Campanula mirabilis*.

Nikolas Michaelovich Albov was on his second journey to the Caucasus. He spotted a plant growing in a rock fissure 91 m (300 ft) above his path, and sent a guide to dig it out. It was the only plant; thorough examination of the neighbourhood revealed no more. When the parcel containing it finally turned up at the Boissier herbarium in Switzerland it was examined minutely but, though it had 100 flowers, there was no seed. A further desperate examination revealed a single capsule, and in it ripe seed.[10] From that, all the stock in Europe was eventually raised.

At Kew it caused a sensation when it formed a pyramid of 300 flowers in 1899: everyone wanted it. But the Rev. Wolley Dod, a practical man, was the first to say that he could not get it to flower. It is monocarpic and difficult, sometimes taking six years to reach flowering size.

The Mountains of Europe

Alphonse de Candolle

Gardeners and botanists alike were excited by the exotic new plants being brought from the Orient. Nearer home, the European Alps had been thoroughly explored by the nineteenth century, and their campanulas recorded. So the time was right in 1830 for the first monograph on the genus, written in Geneva by Alphonse de Candolle.[11] He was born in Paris in 1806. His father Augustin was Professor of Botany at Geneva. At his father's insistence, Alphonse took a law degree, but he never practised; instead, he took over his father's great work, the *Prodromus*,[12] and continued it, writing 26 of the families himself, though even he never finished the work. The dynasty continued with his own son Casimir.

Edmond Boissier

Another great researcher was compiling his magnum opus at the same time in Geneva. This was Edmond Boissier. He was a systematic botanist who had travelled in Asia Minor in 1842 and published descriptions of the new species he found on his journeys. This paved the way for his elaborate *Flora Orientalis* in five octavo volumes which was completed in 1881. It covered a huge area, and was based on his own collecting and that of other botanical travellers.

2 Edmund Boissier 1810–1885

Boissier was that unusual species, a gardening botanist. He grew a large collection of alpines at his home at Valeyres near Geneva. With his friend Henri Correvon, he was one of the founders of the Society for the Protection of Alpine Plants.[13]

Correvon

Henri Correvon, 1854–1939, was a more modern man. He was a conservationist, who had seen 'Peasants of Savoy or Valais bringing baskets full of uncommon species for sale at Geneva market. Classical sites are becoming exhausted.'[14]

He wrote from his own experience a description of campanulas which was serialised in *The Garden* in 1901. It was intended for gardeners, and is full of

3 Henri Correvon

practical hints and personal observation. 'It is a fact that nowhere in the world are the campanulas so much grown and appreciated as in England.'[15] He also recognised one of the problems that would be faced by keen amateurs wanting to grow the alpine species. 'The campanulas being for the most part plants of a strong and healthy constitution easily adapt themselves to new conditions . . . in this way they lose their hairy surface, relax their tissues, increase the size of their leaves and add to the number of their flowers, while contracting their corollas, and so on, so that some species become almost unrecognisable. To such a degree is this the case that one day when visiting Miss Willmott's rich collection at Warley, I found it difficult to recognise certain species of *Campanula*, the actual plants of which had come from my own garden.'[15]

There now started a great fashion for campanulas. The nineteenth century had been a rich period for plant introductions; in the twentieth, eminent horticulturalists took up the work.

Prichard

Maurice Prichard wrote a lavishly illustrated account of campanulas in the *RHS Journal* in 1902. 'Perhaps no family of hardy plant is more generally admired than the Bellflowers or Campanulas, presumably on account of the elegance and informality of their growth, and wonderful freedom of flowering.'[16] He did a fair amount of plant breeding at his Riverslea nursery.

Campanula carpatica 'White Star' which received its Award of Merit in 1905 is still vigorous today. Between 1929 and 1931 six other *Campanula carpatica* cultivars were submitted to the RHS and received Awards of Merit. In 1935 *C. grandis* 'Highcliffe' received its AM after being sent in by Prichard: it is still the strongest variety. And there were many more.

Beddome

By the early nineteenth century the names of many campanulas had attracted a great many synonyms. So in 1907 a retired Indian Army Colonel, Richard Henry Beddome, published an 'Annotated List of the Species of *Campanula*' in the *RHS Journal*. This was intended to sort out many problems for the amateur. 'Nurserymen' catalogues . . . are often very puzzling, so that amateurs wishing to form a good collection are often frustrated, and buy the same plant again and again under different names.'[17] A cry from the heart!

Farrer

Reginald Farrer gave his usual emphatic and salty comments on *Campanula* in *The English Rock Garden*, published in 1918.

4 Reginald Farrer

'*Campanula barbata* is the noble bearded bell of the Alps, and one of their most lovely glories, when its stout campanili of fluffy china-blue wave amid golden arnica in the showering grasses.'[18] Or '*Campanula petraea* is a species of great rarity and great ugliness, from the rocks of Tyrol . . . It is not a biennial, but the gardener sometimes wishes it were, the plant's rarity otherwise protecting it from removal by any reverent hand.'[19] But, for all his jokes, he took *Campanula* very seriously, and his account also wrestled with the synonyms and muddled classification: 'This august race is so vast and complicated that the best thing is to plunge into it at once and go through its serried ranks with care seeing the huge confusion that there reigns and the necessity of weeding the many beautiful sheep from the many goats in the family.'[20]

Crook

The modern standard work on *Campanula* was published in 1951 by H. Clifford Crook: 'Campanulas, their Cultivation and Classification'. The fact that there was an interval of 121 years between it and the previous monograph indicates the size of the task of classification which Clifford Crook undertook: it took him 30 years and extensive travel. He illustrated it with his own black-and-white photographs, mostly taken in the wild, of 216 of the 300 or so species.

This is still the standard work, quoted by both gardeners and botanists, and it is likely to remain so at the moment—though, sadly, out of print. Clifford Crook admired the Greek mountain campanulas, and in 1940 *C. oreadum* received its AM from the RHS. The plant he showed, with its hairy grey leaves, fragile stems and large narrow bells, is a delightful subject for the alpine house.

Bailey

Campanulas are not, in the main, a North American genus, but they had their champion in the United States in Dr Liberty Hyde Bailey, who wrote his 'Garden of Bellflowers in North America' in 1953. It was intended as a horticultural account of the plants which were then available to growers in the U.S. His account is straightforward, and written from a garden 'fully inhabited by Bellflowers' in preparation for the book.

5 Liberty Hyde Bailey

'Perhaps one reason why I have been attracted to the Bellflowers from my youth is because they are not hopelessly confused by hybridisation. There is now a persistent effort to cross everything that is crossable until original lines of singularity are lost, and the natural and distinctive marks of separation have no meaning . . . The best gardener is also the best naturalist.'[21]

Ingwersen

The name Ingwersen is linked to the fortunes of some very good campanulas. Walter Ingwersen introduced *C. poscharskyana* in 1933. It devours space, covering walls and poor soil with its simple star-shaped blue flowers. He has in fact been quoted as saying that it is 'one of the few plants he felt some regrets about having introduced into cultivation'.[22] But it is grown and enjoyed by gardeners across the world who make use of its flowery rampageous temperament. His son Will had an answer: 'Feed the stems and leaves to your childrens' pet rabbits, if they are short of greens—they appear to prefer it to anything else.'[23]

Will Ingwersen VMH introduced *Campanula rotundifolia* 'Olympica' from Washington State, USA, with larger, deep violet bells than the type (AM 1931). The family nursery is famous for its alpines, and they put forward many good species of *Campanula* to the RHS in London. Awards of Merit have been given to *C. laciniata* in 1945, *C. atlantis* in 1952, *C.* 'Birch Hybrid' in 1945, *C. hercegovina* 'Nana' in 1946 and *C. ephesia* in 1956. They have sold and popularised many of the difficult alpine campanulas, but Will Ingwersen returns again and again in his writings to a simple one—*Campanula portenschlagiana*. 'One of the best known and most deservedly popular of all campanulas must be *C. portenschlagiana*. There never was a better plant than this. A measure of its popularity is that, although I find it listed in a catalogue of alpine plants issued in 1903, and that it is to be seen in almost any garden, it remains a best-seller and is in constant and unfailing demand.'[24]

Bloom

Herbaceous campanulas excited Alan Bloom's attention when he was breeding so many herbaceous plants at Bressingham in Norfolk in the 1960s. Dwarf varieties of some hardy but tall stalwarts were one answer to the work of staking, and they looked good in the island beds which he popularised. In 1963 *Campanula glomerata* 'Purple Pixie' received a PC from the RHS. The deep blue and sturdy *C. lactiflora* 'Prichard's Variety' was exhibited by Alan Bloom in 1964 and given an AM. Two years later the well-named *C. lactiflora* 'Pouffe' received an AM, and made the wonderful *C. lactiflora* flowers possible for a small garden. Alan Bloom also received an AM for *C. carpatica* 'Bressingham White' in 1967.

Ever the plantsman, Alan Bloom can always be relied on to spot a good plant. We have him to thank for the re-introduction of *C. trachelium* 'Bernice', which is a double lavender-blue. This could even be the same plant which was described by Parkinson in 1629, which he received from 'friends beyond the seas'.[2] Giving nothing away, Alan Bloom also reported that he received his stock from abroad.

References

1. J. Gerard, *The Herbal or General Historie of Plantes* (1597), p. 448.
2. J. Parkinson, *Paradisi in Sole* (1629), p. 354.
3. J. E. Smith, *Rees' Cyclopedia* (1819–20), under Sibthorp.
4. A. Coats, *The Quest for Plants* (1969), p. 26.
5. R. Farrer, *The English Rock Garden* (1918), p. 205.
6. J. E. Smith, *Letters*, quoted in Coats.
7. F. A. Marschall von Bieberstein, *Flora Taurico-Caucasica* (1808–1819).
8. Curtis' *Botanical Magazine*, 2019.
9. Bretschneider, *Botanical Discoveries* (1898), p. 319.
10. H. Correvon, 'The Genus *Campanula*', *The Garden* (Aug 1901), p. 112.
11. Alphonse de Candolle, *Monographie des Campanulées* (1830).
12. Augustin and Alphonse de Candolle, *Prodromus Systematis Naturalis* (1824–1873).
13. *Gardeners' Chronicle* (Oct 1885), p. 455.
14. H. Correvon and P. Robert, *The Swiss Alpine Flora* (English translation 1911), p. 231.
15. *The Garden* (June 1901), p. 451.
16. M. Prichard, 'The Genus *Campanula*', *RHS Journal* (1902), p. 98.
17. Col. R. H. Beddome, 'An Annotated List of the Species of *Campanula*', *RHS Journal* (1907), p. 196.
18. Farrer, *English Rock Garden*, p. 162.
19. Ibid., p. 187.
20. Ibid., p. 157.
21. L. H. Bailey, *Garden of Bellflowers in North America* (1953), pp. 1, 2.
22. *Gardeners' Chronicle* (Feb 1972), p. 35.
23. Ibid. (Aug 1976), p. 41.
24. Ibid. (1966), p. 353.

Classification

Campanula is the type genus of CAMPANULACEAE, the 'Bellflowers', a family which also includes the other relatively familiar garden plant genera of *Adenophora, Codonopsis, Edraianthus, Platycodon, Symphyandra, Trachelium* and *Wahlenbergia.* Somewhat less well known, perhaps, are *Asyneuma, Cyananthus, Jasione* (well known in the wild by alpine plantsmen), *Michauxia, Phyteuma* and *Ostrowskia. Lobelia* is also closely related, though the irregular arrangement of the corolla here moves it to a separate sub-family, Lobelioidae, in most modern classifications.

With something in the order of 300 species, campanulas are found exclusively in the northern hemisphere, with their epicentre in the Mediterranean lands, particularly Greece and Turkey and their islands, and in the Caucasus. They are strongly represented in the Balkans, and tendrils stretch through Southern Europe, through Spain into North Africa, where the Atlas Mountains in particular harbour a number of species. Further arms stretch through Western to Northern Europe, including the British Isles, to which about half a dozen are probably native, and through Central Europe to Siberia. The Himalaya reveals a miscellany of small, closely related bellflowers which, apart from *C. cashmeriana*, are hardly gardenworthy. Several quite well-known campanulas find their homes in Japan, in Kamchatka, and in the islands, including the Aleutians, stretching up and over into North America, where the genus is not strongly represented, but, if anything, more commonly so in the west.

On a global scale the campanula most commonly found in the wild is the native British 'Harebell' (in Scotland, the 'Bluebell'); this, together with its 30-odd close relatives (if you agree with the splitters!) is found throughout the temperate and subarctic regions of the northern hemisphere, in Europe, Asia and North America. To the gardener this is *Campanula rotundifolia*, the name under which a host of similar plants will be treated in this publication, special mention being made of what, to the gardener, are the more outstanding 'forms', such as *C. linifolia* 'Covadonga' (a name which, alas, has also recently undergone revision).

The flowers of *Campanula* are regular and bisexual, and their parts are normally in fives. The corolla is made up of five petals fused at the base, where the five stamens, though free from it and from each other, are inserted. This is perhaps the place to make a passing reference to the symphyandras, which in general only differ from campanulas in that their stamens are joined at the base, forming a tube. This feature is complicated by the fact that in some campanulas the stamens can also be partly fused, at least in the immature flower, to the point that some botanists have commented that there is little justification other than tradition for keeping the two separated. This said, we admit that we have not included symphyandras here.

The style, which connects the ovary to the stigma, is most often divided in three. In some species it is divided in five, but in each case the ovary has the same number of chambers. The ovary is always inferior in *Campanula*: that is to say, it is a swelling in the part of the flower stem *below* the insertion of the petals and sepals. The seed capsule, again a characteristic of the family, splits by pores when the seed is ripe. These pores may be at the base or the apex; in either case, the ripened capsule is held so that the pores are at the higher end, and consequently the seed only tends to be shed when wind shakes the capsule. This simple device ensures, as far as possible, that seed is carried at least some distance from the plant. In many species the seed is very fine, and can in fact be carried further than one would anticipate.

The calyx, the five green leaf-like sepals below the petals, surrounds the ovary. The shape of these lobes, and the existence or absence of further appendages in the spaces between them, are useful diagnostic features for identifying plants. We make no apology for labouring these features in our plant descriptions for the gardener who wishes to be sure of his names.

Apart from *Campanula vidalii* (syn. *Azorina vidalii*) which is shrubby, all campanulas are herbaceous, most tending to die back to underground rhizomes or carrot-like taproots in winter. The stems bear alternate leaves without small leaflets at their base. A strong family characteristic which associates the family fairly closely with the daisy family, Compositae (Asteraceae), is the presence in the tissues of the milky sap, inulin—a virtue which, although it may protect the plant from the ubiquitous goat to a certain extent, does little enough to protect it from the even more ubiquitous snail and slug, as the gardener soon learns to his cost.

A most interesting feature of campanulas, and this time one more valuable to the gardener, is the pollination method. The anthers ripen within the flower while it is yet an unopened bud. The pollen is shed by them on to the hairs borne on the style. The style only becomes ripe for fertilisation after the opening of the flower. Then its stigmas become receptive to pollen carried by bees seeking nectar in the disc at the base of the flower. In most cases this cross-pollination will be successful, but there is an additional insurance mechanism in the unsuccessful case; here the stigmas gradually curl over and down until their adhesive tips will pick up pollen from their own style where it was left by the now withered stamens. The fail-safe mechanism greatly increases the chance of successful fertilisation and consequent seed formation.

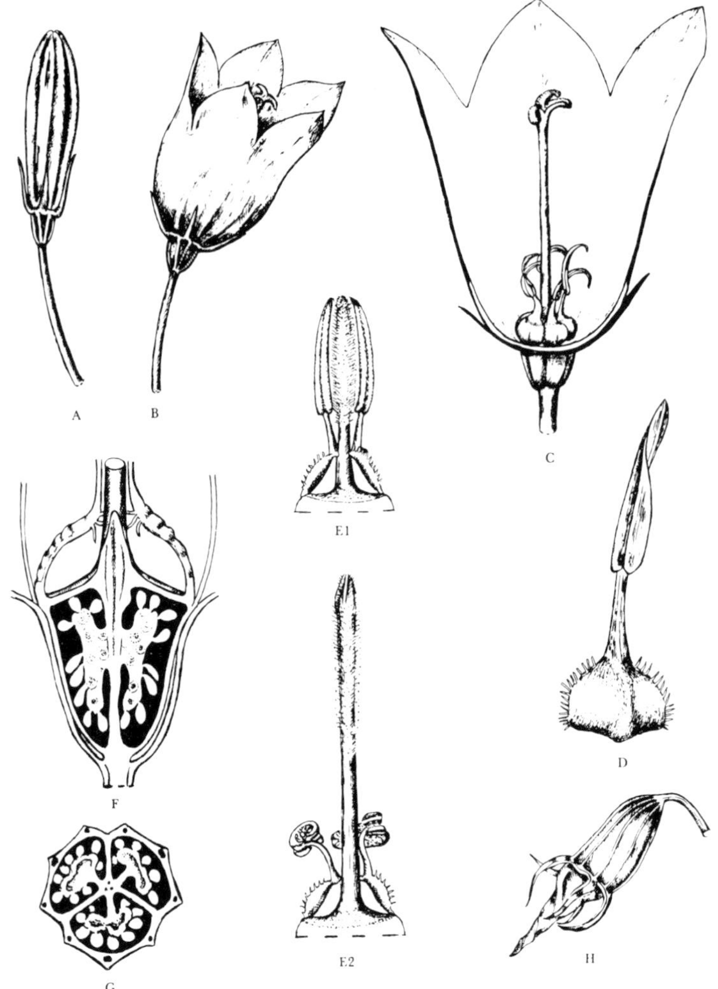

6 *Campanula rotundifolia*: the floral structure

A A flower-bud, showing the corolla surrounded by the five linear lobes of the calyx. The ribbed calyx-tube is joined to the inferior ovary. Bud 12 × 3 mm.

B The regular, bisexual flower at time of flowering, shown as erect though it is nodding on the plant. The bell-shaped corolla has five lobes (petals) that are shorter than the tube (cleft, typically, to one-third of the corolla.)

C Section showing half of corolla, with stamen and style. The 5 stamens alternate with the corolla lobes, and the expanded bases of the filaments form a dome over the nectar-secreting disc at the base of the style. The anthers have shrivelled, and the three branches of the style have commenced to separate in order to expose the stigmatic surface to which pollen will eventually adhere for pollination to be effected.

D A complete stamen. The anther is two-celled, and splits along its length to shed the pollen within the flower—a characteristic of (potentially) self-pollinating flowers; in campanulas the pollen is shed on to the style.

E1 The style and two of the five stamens as they appear in the flower bud. At this stage the style is the same length as the stamens, and the anthers are able to deposit their pollen on to the hairy upper part of the style.

E2 The style and two of the stamens in the young flower. The style has lengthened, and the pollen-covered portion extends well beyond the stamens, which have now begun to wither. (The next stage of development is shown at (C).)

F Vertical section through the inferior ovary. The numerous ovules (each of which may form a seed) are borne on a structure radiating from the central axis. Above the ovary is the nectar-secreting disc protected by the expanded bases of the filaments.

G Cross-section of ovary.

H The fruit, before shedding of seed. The withered corolla and calyx enclose the capsule, which opens by pores to shed the seeds.

From *100 Families of Flowering Plants*, by Michael Hickey and C. J. King, Cambridge University Press.

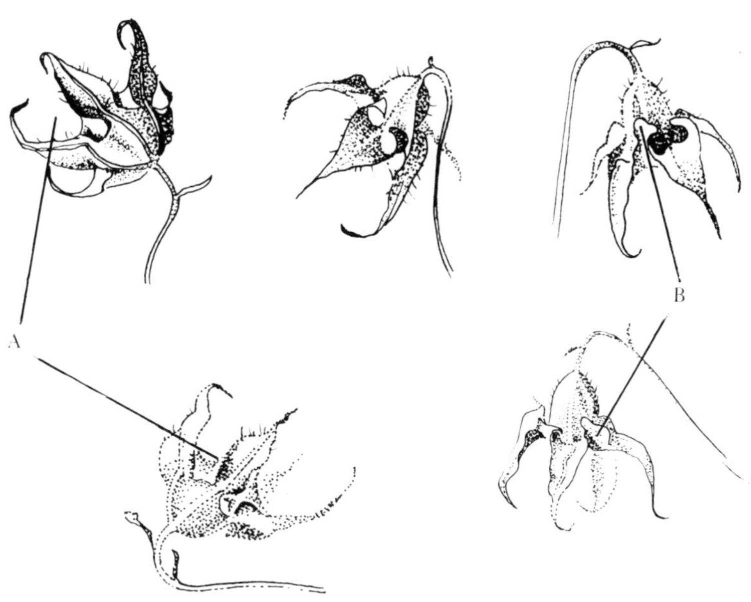

7 *Campanula punctata*: calyx with flower shrivelled. A. sinus, B. appendage

From the cultivator's point of view, this in general means an abundance of seed set. In those species where there would appear to be a measure of self-sterility—this applies particularly to the high alpines—there is still usually a small amount of viable seed to be obtained, albeit with much patience, good eyesight and agile fingers.

Hybrids are known, though rare in nature; in cultivation they are not as numerous as the size of the genus would suggest, Even then the principal perpetrators are few: *Cc. carpatica, cochleariifolia* and *rotundifolia*; and, to a lesser extent, *Cc. isophylla, punctata* and *pyramidalis*. These give rise to some of the most popular cultivars, propagation of which is, of course, confined to vegetative means.

The prevailing colour of the bellflowers, it is hardly needful to say, is blue, from a deep violet to the very palest milky blue. White variants occur frequently. There is a red element in the inheritance which, in some species and a few notable cultivars, shows itself in soft pink colorations. One or two species reveal a yellow inclination, but except for the tender annual, *C. sulphurea*, an attractive little plant from the Lebanon, this yellow manifests itself in a rather shabby-looking straw shade. This applies to the commonest of this colour, *C. thyrsoides*, which is perhaps more interesting than beautiful; its subspecies, *carniolica*, can be really yellow.

In the genus are annuals, biennials, monocarpics and perennials, which can be both long- or short-lived. In general the annuals are not a very interesting lot, and almost without exception are not met with in cultivation, which is neither surprising nor likely to change. The biennials include, of course, the Canterbury Bell, which has now rather fallen out of favour for no detectable reason, except that, being considered a native of Britain, it is perhaps considered too ordinary. A recent increasing interest in wildflowers could change this. The monocarpic campanulas—that is, those which may last several years before flowering, after which they die—are numerous, especially from the Mediterranean region. Most of these are a lot hardier than is generally realised, and for bonuses have a decorative leaf-rosette while maturing, and also leave copious seed for propagation purposes. The perennials form the backbone of the cultivated species, showing a remarkable range of forms, dimensions and degrees of difficulty—or ease—in cultivation.

Campanulas vary in size from ground-hugging alpines barely 2.5 cm (1 in) high, rambling attractively over the rocks or scree in which they grow, to majestic pyramids, 2 m (6 ft) or more in height which rival their illustrious cousin *Ostrowskia magnifica*, a superb but difficult plant rarely seen in gardens.

As one expects of any family, botanical, zoological—or human!—certain characters and tendencies crop up repeatedly to a greater or lesser extent. Roots, whether rhizomes or taproots, tend to be woody, thick and coarse. Basal leaves frequently wither near or at flowering, and tend to be on fairly long stalks; stem leaves are on shorter ones or completely stalkless. Most leaves are

8 *Campanula punctata* 'Alba'

9 *Campanula excisa*: flower

10 *Campanula cenisia*

either oval, heart- or lance-shaped in varying degrees; many are notched or toothed.

The family has another characteristic, which can be very confusing. The plants' size and outline can change very considerably depending on where they are grown. The mountain plants especially will be much larger and floppier in soft lowland conditions. A number of writers have commented on this, and said that sometimes a plant can be unrecognisable at first glance. Such features as leaf size, length of stalk, flower size, pendent or upright habit, are all variable when conditions change.

The botanical features of the flower have been summarised, and, apart from the fixed feature of their petals being fused at the base, the flower presents as a bell, as a trumpet, as a funnel, or as a star—also, *vide C. carpatica*, as a salver or saucer. In the garden flowers may be doubled in the most popular cultivars, when they can present as cup-and-saucer, cup-in-cup or as multiples to rival the cabbage-rose. Such doubles usually demand more care for them to give of their best.

Campanulas are high-summer flowering plants, essential ingredients of the summer border, and indeed, of the rock-garden. One has only to read William Robinson, Gertrude Jekyll and Reginald Farrer to realise that the garden would be immeasurably poorer without them even when they are not consciously given pride of place.

11 *Campanula lasiocarpa*

The Border Plants

Cultivation

The cultivation of campanulas, particularly those we grow in the open border, presents no special difficulties. The species have no strong likes or dislikes, and will tolerate most reasonable garden conditions. To have them give of their best, however, the sensitive gardener (is this what 'green-fingers' is about?) will realise that just as gardeners give of their best, like anyone else, when they are free to express their own preferences, so will their plants when the same rule of (green) thumb is applied.

Campanulas prefer sun for at least part of the day, and some for most or all of it. They will, however, tolerate quite heavy shade if inevitable. Lime in the soil is a positive preference except in the case of a small minority—which do not concern us here. Good drainage is essential; a low-lying spot which is waterlogged in winter for any length of time is guaranteed to rot taproot or rhizome beyond recovery. No doubt this applies to most garden plants other than bog-plants.

Many campanulas grow on quite impoverished soil in nature, and in habitats where competition from other plants—especially grasses—is strong. Although in such circumstances they give a good account of themselves, they are never of the stature or floriferousness we look for in the garden, and of which they are more than capable. Examples are *C. persicifolia* as it grows in the high country meadows of the Pyrenees, where it is hard put to grow to 30 cm (1 ft), and bears but a few flowers on its meagre stem, and *C. glomerata* on the thin soils of the South Downs of England, where the 'cluster' of the Clustered Bellflower is but a poor sight. In his highly respected monograph Clifford Crook depicted the former in the wild, and it is scarcely recognisable.

Many of the campanulas are great travellers! A few need to be watched on this account, especially if the soil be sandy or peaty; not many are positively menacing, and it is not really a problem to deal with them as necessary, but it does betray a liking for new ground. An occasional lifting, with some division of rootstock, and replanting will reinvigorate even the most static stay-at-home.

If there be a 'florist's' flower in the genus, it is *C. persicifolia*. This reacts well to being moved on every year or two. The more finicky doubles like to be moved every year, and require feeding in addition. For the generality of border bellflowers a yearly sprinkle of bonemeal pricked into the surface of the soil will amply suffice.

After flowering the stems may be cut back to prevent any indiscriminately self-sown seedlings arising; a modest second flowering may result; or seed may be saved—it is usually copious. If it is to be saved, the warning of, it is believed, Will Ingwersen, is recalled: 'Many a watch has been kept at the front door while the seed has slipped out of the back unnoticed.' Timing is the watchword.

Propagation

Apart from saved seed, there is not a wide selection of species available in commerce, and the chief source must be the seed exchanges of the specialist societies; over the past few years nearly 250 variants of *Campanula* have been offered. It is possible to recover more than the cost of the annual subscription to such societies on the value of the seed alone, apart from other benefits including the shared expertise of other growers in their periodicals. Addresses of these societies may be found on p. 138.

All campanulas are easily grown from seed, which may be sown in late winter or early spring in trays or pots of a good compost; a John Innes based compost* is considered best, with the addition of half the volume of extra grit. The seed should be sown thinly on the surface, and covered with only the finest layer of sharp sand or fine grit. This will discourage liverworts, but should be thin, as light is beneficial to germination, so the trays should not be covered with paper, etc. As the seed is fine, watering is best done by soaking the tray in water before standing out. Frost will not harm the seed, even those species growing towards the southern end of their natural range, and may well hasten germination by breaking dormancy, but the trays should, of course, be sheltered from heavy rain, wind, birds, etc.: a cold frame is hard to better. After germination the seedlings may be brought into some warmer spot so that they may be kept moving, and pricking out should be done when they are quite small, with four or so true leaves, when they will quickly establish; but no harm will be done if delay is unavoidable. They can be potted on if necessary, and then planted out when the soil has had the opportunity to warm up and a strong-growing rosette has been formed. Any special treatment will be given under the individual species.

There are those who prefer, or find it more convenient to use, soil-less composts. Germination and early development of the seedling in this case may be very good, but we find that problems tend to start when it is desired to 'wean' plants so started into open ground. The lush top-growth makes an attractive-looking plant, but the soft root growth receives something of a shock

* John Innes is seven parts sterilised loam, three parts peat, two parts sharp sand, with limestone and fertiliser according to the plant's stage of development.

when it has to tackle the rigours of a garden loam, however well tilled; hence the advice given with many commerical plants to mix an abundance of peat in with the soil into which the plant is being set out. In our opinion a soil-grown plant is more enabled to adapt and, in the long run, is the quicker to establish. The reader may conclude that this is a matter of prejudice, but it is established on personal experience. However, 'the devil you know . . .', the properties of which are best known to the individual grower, will in each case give best results.

We have mentioned that frost access to the seed-tray will be beneficial; this applies to the vast majority of campanulas, including the Greek and Turkish ones, and many from the Mediterranean. When seed arrives, as so often it does, too late in the year to catch such frosts, it may be mixed with a small amount of moistened, very fine sand, and placed in a polythene bag in a refrigerator for a few weeks as artificial vernalisation; but particular care needs to be taken with the fine seed of campanulas, lest the final sowing be too deeply covered.

We have found that *Campanula* seed, placed in a well-sealed plastic box with a little silica gel, will keep viable in a refrigerator for a number of years, and we make a regular practice of storing in this manner.

Vegetative propagation from established plants is best undertaken in the spring, as new growth is getting going. Clumps can be divided quite drastically, when any piece with a root or vestige of root may be placed in a gritty compost until it is well rooted, and then planted out as desired. Additional feeding, except in the case of some of the *C. persicifolia* cultivars, especially the doubles, is unnecessary.

Cuttings may also be taken, again preferably in the spring when new growth is strong. A peat-and-sand mix, or sand and vermiculite or other proprietary cuttings mix, for preference in a propagator where watering is easily controlled, are suitable media. Some bottom heat may hasten root formation, but is not essential. A clean cutting, with lower leaves removed should be placed in the medium, watered in, and taken out for potting when sound roots, betrayed by firmness of hold in the propagator and evidence of fresh green growth, are established. As campanulas are not, in the main, hungry plants, there is little need for transfer until good root formatioon is made.

Pests and diseases

As a genus, *Campanula* is outstandingly trouble-free in cultivation. The primary enemy is the slug, but this as a general rule is only troublesome with some of the more succulent (and, of course, rare and valuable!) smaller species. In particular, a slug of 3 mm in length can decimate a pan of seedlings in a night, or eat out the growing bud of a delicate plant as rapidly. Various authorities place faith in various slug-baits, whilst others refute them as stoutly. Precautions such as the placing of plants, standing seed-trays on the very sharpest of sands (silver-sand, for example) seem as effective as any. Aluminium sulphate (sold as Slug and Snail Powder by Fertosan) in powder or solution around the plants at risk has proved useful; this has the advantage that it is not toxic to pets, birds or earthworms, but sprinkling on the foliage should be avoided.

The species particularly at risk are indicated in the descriptive texts, but it must be emphasised that the vast majority are not in any way menaced, and this includes virtually all the border plants.

Aphids rarely trouble campanulas, even when they are abundant on nearby plantings.

Rusts have been described in the literature, but we have not often seen them in 20-odd years of growing campanulas. If they should occur, a general book of plant pests should be consulted (*Collins Guide to the Pests, Diseases and Disorders of Garden Plants,* Buczacki and Harris, is readily available).

In the same general way damping off of seedlings may be encountered; though not commonly. Precautions such as sowing seed thinly and not over-watering will usually be quite adequate.

Descriptions

The following is a description of the border campanulas in alphabetical order, but we have included some variants which can also be grown in a rock-garden, but which are vigorous enough to make useful front-line plants in the border. Our cut-off point is about 30 cm (1 ft), but it will be realised that the eventual size of any plant will depend on both cultivar and the conditions under which it is grown. Sizes given, therefore, are very approximate.

The descriptions are intended to be non-technical, but to give enough information for readers to be able to identify most of the plants they meet. A few common scientific terms have been used because without them the text would be impossibly long: they are explained in a glossary at the end. Readers who need a more precise and detailed description are referred to the Bibliography, and the various regional *Floras*, where complete descriptions may be found.

There are over 300 species of *Campanula*, many of them only known to Eastern European botanists. We hope that some of these plants may eventually become available to us in the West.

In the following descriptions of individual plants, the first figure indicates height, the second the spread, of a mature plant. In the absence of a second figure it is understood that the actual spread will depend on age, soil conditions, etc. This applies in the cases of most which form spreading mats. Also indicated are the main flower colours and the visual month(s) of flowering.

Each of the following accounts includes the most recently accepted nomenclature, together with synonyms which may be come across, especially in older literature. It is hoped that this will be a help to those who like precision, and easily ignored by those who find it pedantic.

C. ABIETINA Grisebach. 60 cm × 30 cm Violet July

Rosettes of oval leaves, wider at the base, or broadly lance-shaped leaves, are 5 cm long on short petioles. They give rise to a number of fine stems bearing longer, narrower and stemless leaves, often more abundant at the tops. The stems and leaves are hairless. The flower is a wide open bell, the petals spreading from the centre like the spokes of a wheel. They are about 2–3 cm across and of a colour variously described as light violet or reddish-violet. The calyx, about one-third the length of the corolla lobes, is awl-shaped, smooth, not toothed or cut, with no appendages.

C. abietina occurs in stony mountain grassland in the Carpathian and North Balkan mountains; also in the Caucasus.

Flora Europaea makes *C. abietina* a subspecies of *C. patula*, a somewhat coarser biennial plant of lower elevations throughout Europe. Whereas *C. patula* is biennial, *C. abietina* is markedly perennial, even if not a long-lived one. A clear sign of this is that it throws stolons, but from the point of view of the gardener not menacingly so, as with certain species of *Campanula*.

C. abietina is an attractive, if somewhat neglected, plant eminently suited for the front of the border where its relative delicacy is appreciated. It sets seed well, and may be easily propagated from winter or early spring sowing; or alternatively, divided annually, when it appreciates a rich but well-drained soil. The plant received an AM in 1891.

SYNONYMS: *C. patula* L. ssp. *abietina* (Grisebach) Simonkai; *C. vajdae* Penzes.

C. ALLIARIIFOLIA Willd. 60 cm × 45 cm White July/Oct

A tuft of heart-shaped leaves, on stalks up to 20 cm long, forms the base of *Campanula alliariifolia*. This throws up a clump of leafy stems, occasionally branching and bearing a one-sided raceme of flowers up to 5 cm in length. They are narrowly bell-shaped and white or creamy-white in colour, though the wild type is given as purplish-white, a shade rare in cultivation. The corolla segments are about one-fifth of the petal's entire length, and slightly reflexed. The calyx segments are about one-third of the corolla length, with softly hairy circular appendages. The three part style does not protrude beyond the corolla.

This is not the most elegant of campanulas, but it is hardy, soundly perennial and long-lasting. The basal clump of leaves is fairly coarse, but not undecorative in itself; these leaves are felted above and quite woolly beneath, the leaf-margin slightly toothed and looking characteristically somewhat crumpled when young. It is not at all fussy as to soil, but does not need a rich one. It tends in our experience to stand up better in a poor soil, when grown in exposed and windy positions. It also gives a better effect when a clump of three or five are placed close together.

C. alliariifolia occurs in Anatolia, Turkey, and in the Caucasus, growing in open scrub and conifer forest, occasionally on cliffs, but frequently on steep banks; in fact it appears to its best advantage if it can be placed at the top of a bank. The flowering period is long, and it will repeat well if dead-headed.

Propagation is easiest by seed—it self-sows modestly; seed is set abundantly. Division in spring is also easy.

C. a. 'Ivory Bells'

The form most often offered; creamy-white. No special authority has been found for the name; some use it as a vernacular name for the species rather than a cultivar designation.

C. a. 'Flore Pleno'

A double, not often seen and said to be not particularly attractive.

SYNONYMS: *C. lamiifolia* Adam.; *C. macrophylla* Sims.

C. ochroleuca Kem. Nath.

This species may also be mentioned here; it is by some considered as a subspecies of *C. alliariifolia*. It occurs in Transcaucasia and is, as the name betrays, a creamy-white in colour, which in a good form is attractive. The chief difference is that the style extends beyond the corolla. The two are, of course, inter-fertile, and it is more than likely that the material in cultivation is of entirely haphazard race due to hybridisation.

C. BETULIFOLIA C. Koch 30 cm × 40 cm White/pink June/Sept

Some fine forms of this plant have been collected as seed in the wild in recent years, and it is proving itself an adaptable and versatile subject which can be grown in the front of the border, on the rock-garden, in sinks or in pots with equal success. It justifies a prominent foreground position where its charming delicate appearance can be appreciated.

A tuft of thick and glossy basal leaves arises from a perennial woody rhizome. These leaves are pale to mid-green, somewhat bluish, long heart- to wedge-shaped, with wavy indented margins. The petioles are somewhat longer than the leaf-blade. There arise from this tuft numerous wavy stems bearing more or less oval, toothed leaves, on long stalks. The plant's name is a reference to their 'birch-like' appearance. These stems are branched above, each branch bearing up to five flowers in a loose cluster. The buds are characteristically wine-red, and open up to a relatively large and well-shaped flower, white or with a pinkish tinge, often heavy enough to make the branches somewhat floppy. In the best garden forms—it is a very variable plant—the corolla is a good bell-shape and the stems sturdy enough to hold the flowering branches upright above the foliage. The calyx lobes are triangular with short appendages.

This is a soundly hardy perennial which was given an AM in 1937; in the plant then exhibited the flowers were of a pale pink.

Cultivation presents no difficulty, so long as there is good drainage in the soil habitat; this should be moist but not too rich, as this tends to give excessive growth of leaf at the expense of flowers, which also become hidden in the foliage. It is a plant from varying elevations in Turkey, Armenia and the Caucasus, and the higher forms are reliably hardy provided that drainage is good and roots do not freeze for too long in frosty weather.

C. finitima

This form falls within the above description, and there is little doubt that in common circulation there is no difference. In the original descriptions *C. finitima* was given as having more cup-shaped flowers than the narrower bells of *C. betulifolia*.

Similar species are *C. troegerae* and *C. choruhensis*, both quite recently described, and also occurring in Turkish and Russian Armenia. All three are grouped botanically in the Symphyandriformes Section of Campanula—approaching *Symphyandra* in that the anthers are sometimes fused into a tube around the style, as in that genus; but here only in the young flower.

SYNONYMS: *C. finitima* Fomin; *Symphyandra finitima* Fomin.

C. BONONIENSIS L. 76 cm × 50 cm Blue July

This is a greyish plant whose leaves have short soft hairs. It has a basal rosette of stalked, heart-shaped leaves, about 7 cm long. They throw up a stem bearing leaves which become progressively shorter towards the tip; these are stalkless and partly clasp the stem at the leaf axils. There is a long, spike-like branch of smallish flowers, clustered in twos or threes on short pedicels. The flowers are smooth, funnel-shaped, the lobes divided to about one-third their length, and are blue. The lance-shaped calyx lobes are much shorter than the petals, and very bristly. The style is in three parts and about the same length as the corolla. The flowers, and the subsequent seed capsule, are pendent.

C. bononiensis occurs over a widespread area and conditions; Central and Eastern Europe, Iran, Caucasus and Western Siberia, at forest margins, scrub and rocky ground. It was first described from the Bologna area of Italy, hence the name, Bononia being the Latin form.

This is not a delicate plant, as its origins might suggest, it is not hard to suit, accepting virtually any soil, shade or sun. Crook condemns it to the more extensive wild garden, and also assures us that it is biennial. *Flora Europaea* and

Flora of Turkey give it as perennial. It would seem to us that some clones are more permanent than others, and also a rich soil tends to spoil it away quicker. Whilst it is not an aristocrat of the border, a good cluster can be effective. Graham Stuart Thomas refers to it as a refined plant with graceful spires; a good form is just that.

Seed is set freely, and germinates well (we suspect that light is necessary for this so it should not be covered; a sprinkling of fine grit suffices to protect the seed when the pan is put in the open to germinate). A good form may be divided in spring as growth restarts.

SYNONYM: *C. ruthenica* Bieb.

C. 'BURGHALTII' 60 cm × 30 cm Purple July/Sept

This, a very old hybrid whose origins and history are lost, is generally accepted to be a cross between *C. punctata* and *C. latifolia*. It has more of the form of its seed parent, that is, *C. punctata,* than of the pollen parent, *C. latifolia*. It is, in fact, the reverse cross to *C.* 'Van Houttei' which see.

The basal leaves are heart-shaped, on stalks as long as the leaf-blades, and very slightly winged. Progressively up the stem, the leaves become more oval, wider at the base, lance-shaped and also stalkless. The upper surface of the leaves is smooth, but there are short bristly hairs on the veins of the underside, and also on the stems. The stems are somewhat branched, and the flowers are born terminally, and in each leaf axil. These stem leaves are usually very pointed and upward-facing like ears. The flowers are amethyst-purple in bud. About two-thirds of the way from the tip of the bud is a little hook in each lobe. When the flowers open this marks the point at which each lobe is split. They are slightly hairy within and dove grey to light mauve when fully expanded. However, they do not open widely, and form a long narrow bell some 7–10 cm long. The spent flowers change colour again to a true blue as they die. The calyx lobes are long-pointed with turned back triangular appendages. The three part style does not protrude.

Evidence of hybrid origin is the fact that seed is never set, and the only means of increase is by division in spring, or by taking fresh spring growths from the rootstock which are easily established.

This is a sought-after plant which is always in short supply, as it does not spread anything like as vigorously as the *punctata* parent. It is said of it that it has a slightly running rootstock; there is little evidence of this in a heavy soil, but in moist light soil this quality might show itself. In any case, *C.* 'Burghaltii' prefers some moisture, and is best in part shade, where, when happy, it will flower for months continuously. It may need some support. John Raven, in *A Botanist's Garden*, touches on this; 'instead of lifting its many spires straight upwards, . . . it sprays then gracefully out in all directions so that it is

12 Campanula 'Burghaltii'

appreciably wider than tall—and its large pendent flowers are of a unique and soft shade that is quite as much pale grey as blue.'[1]

C. 'Burghaltii' is a long-lasting perennial which is soundly hardy. Like many another campanula it is completely deciduous, and its position should be well marked in the border before ambitious winter or early spring tidying up is undertaken.

C. CARPATICA Jacq. to 45 cm × 45 cm Blue, White Summer

Campanula carpatica has suffered some neglect in the past by falling into that ill-defined no-man's-land category between the rock-garden and the border where the dedicated devotee of each has left its promotion to the other. The least disturbed by such treatment or lack of it, however, is the plant itself: in its carefree resilience it will be equally happy wherever it is grown.

Coming from the Carpathian Mountains as the name suggests (and to be carefully distinguished from the very different *C. carpatha* from the Greek island of Karpathos), *C. carpatica* has a perennial root stock which throws up a tuft, and, when well established, a clump, of bright green roundish leaves which, as with a number of other species, disappear as the stems grow, to be replaced by more oval to triangular ones on long petioles. These are repeated up the stems, becoming more heart-shaped, the petioles gradually diminishing. Toothing of the leaf-margin varies, as does the amount of branching of the flower-stems. The flowers are borne terminally. Because this species is very variable it has, along with *C. persicifolia*, attracted attention from the florists in the past, with the result that a large number of clones have been selected and named: thus form and size of plant, shape and colour of flower are consequently extremely variable. In the wild type-plant the flower is a pale narrow bell, but in cultivation colour is between purple and white, and the shape anything from funnel to cup to umbrella to open salver. The short wide corolla lobes are sometimes rounded, and at others bear star-shaped points; the flower is held boldly to the sun. The calyx teeth are lance-shaped, and there are no appendages. The style divides into three.

C. carpatica is a sun-lover, and ideal for the front of the border. Its cultivation is in general of the easiest, and it will thrive where it is free from winter waterlogging. It is very free-flowering, and if 'shorn' as the flowers fade will give a second crop, almost as good, later. In fact it is one of the longest-flowering campanulas. Seed sets abundantly, and the more vigorous sorts self-sow copiously; excess seedlings may be easily removed. Otherwise this is not an invasive species, but one which with a minimum of attention will give a minimum of trouble and a maximum of effect in the garden.

C. turbinata

A form variously described as *C. turbinata* or *C. carpatica* subsp. *turbinata* is said to be dwarfer and with unbranched stems, the flowers being the shape of spinning-tops ('turbinata'). The influence of this form is seen in some of the named cultivars.

More than a dozen cultivars have over the years received Awards of Merit, incidentally beating the record of any other campanula. A by no means exhaustive search of literature and lists will quickly turn up as many as 50 cultivar names of varying authority and worth, but many of these of impeccable derivation and pedigree have been lost, some of the AM winners among them. It should of course be noted that no named cultivar may be had true from seed; when offered, such are at best only strains, however little variation they show. Propagation may be by seed, or by cuttings in spring as growth restarts. Only by vegetative propagation may the cultivars be had true to name. This, of course, applies to any species of any genus.

When the total number of species of *Campanula* is taken into account, there are relatively few hybrids about—the operative word being 'relatively'. In the wild they are virtually unknown, but are commoner under garden conditions. Of those species which do tend to hybridise, *C. carpatica* is certainly one of the more promiscuous, most of the resulting progeny being on the rock-garden scale. It must be said that in the list of cultivars given here there will be a few whose exact origins are now untraceable, though an intelligent guess may be hazarded. The following are among the more readily available cultivars in the UK. Some of the more traceable hybrids will be given treatment in the alpine section.

C. c. 'Blue Clips'

China blue cups on a 15–20 cm hummock; this is a very reliable plant which in fact comes very nearly true from seed.

C. c. 'Blue Moonlight'

Appropriately named very pale blue open saucers in a dense hummock; very free-flowering, 20–25 cm.

C. c. 'Bressingham White'

Selected by Alan Bloom, this has open white flowers with a blue-green trace at the base of the petals. 20 cm. Received an AM in 1967.

C. c. 'Chewton Joy'

Received an AM in 1929, and is still going strong. It is about 20 cm, free- and late-flowering; open cups of China blue with paler centre to the cup.

C. c. 'Hannah'

Though the flowers are relatively small they are borne so abundantly and at a time when greatly appreciated—late summer, and are of such a good white that this cultivar, which is almost certainly of *turbinata* parentage, is outstanding. A cushion 20 × 30 cm when established.

C. c. 'Isabel'

This is one of the many successful campanulas introduced by Prichards of Highcliffe, a seedling from *C. c.* 'Riverslea' in 1905, and therefore again of *turbinata* origins. It is a deep blue flat flower, described originally as purple-blue, about 25 cm in height, and its longevity in cultivation is evidence of a sound constitution. AM 1904.

C. c. 'Karl Foerster'

20 cm. Large, open, deep cobalt blue saucers, this is named after the respected German plantsman, himself responsible for many good plant introductions. It

flowers early and has been forced for the Chelsea Flower Show, when it looks what it thus becomes, a house-plant. It is, however, robust and long-flowering and, like many listed here, fitted for the front of the border as much as for a pot.

C. c. 'Riverslea'
Prichards introduced this, and it is almost certainly the 'Giant', alias 'Riverslea Giant' which obtained an AM when shown in 1931. It was described as a little over 30 cm with large flat deep purple-blue flowers 5 cm in diameter. It is not now easy to find, a fate shared with many another worthy cultivar, but it is still available.

C. c. 'Snowdrift'
This is a good clean white flower of medium size; a slow-growing plant, compact in form.

C. c. 'Wheatley Violet'
This is a low-growing *carpatica*, one of the deepest violet and showing the tendency to hairiness of *turbinata*. We understand that this was a Valerie Finnis introduction.

C. c. 'White Clips'
Like its blue counterpart, an introduction from the German Benary nursery, and a strain which comes fairly true from seed; it bears a clear white flower held some 25 high on a tidy plant.

C. c. 'White Star'
Received an AM in 1905; again from the Prichard nursery at Riverslea. It is of open form in flower, with characteristic pointed tips to the star-shaped corolla lobes, satiny white in texture, held about 30 cm high. This is no longer easy to find.

Many other cultivars are now hard, if not impossible, to find, but among them the following would be worth the tracking down:

'Big Ben'
'Blue Bonnet' (AM 1968)
'Blue Star'
'China Cup'
'Clarence Elliott' (AM 1937)
'Claribel'
'Convexity' (HC)
'Craven Bells'
'Ditton Blue'
'Far Forest' (HC)
'Glacier Blue'
'Grandiflora' (AM 1967)
'Harvest Moon'
'Jewel'
'Lilliput'
'Little Gem'
'Loddon Bell'
'Loddon Fairy' (AM 1967)
'Queen of Somerville'
'The Pearl' (AM 1967)

'Wedgwood'
'White Convexity' (AM 1947)
'White Gem'

Without being, or wishing to appear, cynical, we have to say that our experience has taught us that many plants offered under any of these names could need careful checking for authenticity.

C. COLLINA Bieb. 35 cm Deep blue June

The campanula made Reginald Farrer wax lyrical—no hard task for him. 'One of the most gorgeous campanulas we have, with tufts of downy foliage scallop-edged and oblong heart-shaped on longish foot-stalks, and then the graceful foot-high stems, gracefully carrying magnificent big bells of imperial purple, satiny and brilliant . . .'[2]

This is a perennial with a slowly creeping rhizome. The basal leaves are just as above, and the erect stems bear similar leaves which are gradually reduced up the stem to stemless bracts. The nodding flowers are carried in a one-sided cluster, up to five in number, but sometimes only single. The petals are dark purple to blue-violet, and in the form of a longish, but full, pendulous bell, the segments of which are split to half and curl well back. The calyx lobes are lance-shaped, not reflexed, and are without appendages. The three part style does not protrude beyond the petals.

Campanula collina does not appear to be fussy, and will grow equally well in limy or lime-free soil, though it is reported as coming from mainly granitic soils. It grows wild only in Armenia. In cultivation it does like a moister soil than is suggested by the pastures and rocky soils of its origin (it does not grow in rocks), and it will quickly succumb if the soil dries out in the garden. In a good, open loam it will thrive and spread, though never menacingly. It takes neither the eye nor the pen of a Farrer to appreciate the gorgeousness of a mat of it in flower, though its flowering is not long. It is one of the campanulas which, being completely deciduous, leaves no trace of its winter hiding-place, does not revel in being dug about while resting, but does appreciate a good mulch both in recognition and in encouragement. Propagation is easy by division in spring, or by seed.

This plant figures in the *Botanical Magazine* at t. 927 in 1806. It is a very variable species; several naturally occurring varieties have been described, and that depicted in the above was given as the 'Major' form, which does not now appear to be available.

C. DIVARICATA Michx. 90 cm Pale blue July/Aug

It appears strange that, relatively speaking, so few North American plants are common in cultivation, when in the wild there is such abundance, and when those from countries which have been less systematically botanised are so well known. Perhaps it is that the American way of life has not in the past encouraged the widely dispersed love of gardening known in Britain and parts of Europe; and on the gardening, as opposed to the botanical, level, reliance there has been placed on European tastes and literature, with a consequent neglect of American natives. But is the grass always greener over the fence? Whatever the answer, we are now learning, none too soon, that the great American continent has a lot to offer the lover of flowers.

Clifford Crook in his monograph described this species as being of considerable charm. Whilst North America is not strong in campanulas, this one could justifiably be depicted as its best. It is a perennial from the Southern Appalachian Mountains from Maryland south to Georgia, occurring on stony banks in both sun and shade.

Campanula divaricata forms clusters of rosettes of dark green oblong or lance-shaped leaves which taper at both ends to a slender point. Smooth slender upright stems with few or no leaves branch repeatedly and bear towards their tips panicles of numerous small flowers (1 cm long or less). They are pendent bells of a pale blue shade, shaped individually not unlike those of some forms of *C. cochleariifolia*, and reminiscent of an old-fashioned mob-cap. The three part style extends beyond the petals by as much as the petal length, giving a strangely characteristic profile, for the long-styled campanulas usually have a circular or star-shaped corolla. The calyx, like the rest of the plant, is smooth, with narrow pointed lobes that have no appendages.

This plant presents no difficulty in cultivation, and generally sets abundant seed. With it, we break our general rule of recording here campanulas which are readily obtainable or nearly so. The plant is not yet available, but should be!

C. flexuosa Michx.
This form has been described, and appears to be but a compacter alpine form of the type; to some this would be even more desirable.

C. FORMANEKIANA Degen & Doerfler 30 cm White July/Aug

There are a number of campanulas from Greece and the Balkans which, because of their woolly 'finish', may give the impression that they are tender or fragile. In fact, it is not as simple as this, but frequently the case that a little trouble taken in siting a plant wisely will be amply repaid.

These campanulas from the warmer areas of Southern Europe are in the great majority of cases not afraid of the cold in itself, even of prolonged freezing temperatures, but will suffer from alternate freezing and thawing if it is their roots which are frozen. Place *C. formanekiana* well up in a wall or well-drained rockery so that its roots are dry, and will not be frozen, and it will give a wonderful display.

A stout taproot forms a silver-grey rosette of particularly neatly arrayed

13 *Campanula formanekiana*

leaves. These are oval with a wider heart-shaped base, and slightly rounded at the apex. The petioles are long and winged. Normally, an upright central stem is accompanied by a number of lower, decumbent branches. The stem leaves have shorter and shorter petioles up the stem, and are also smaller. The flowers are large 'Canterbury Bells', about 4 cm long, in white, pale blue or pale pink with a three part style. They are held erect and have large triangular calyx segments without any appendages.

The plant is a Balkan species which grows in the mountains of Macedonia. It was first collected by Formanek who published a description of it in 1895 and called it *C. cinerea*. This name had to be dropped when it was again collected and described in 1897 by Dorfler, as it had already been used for a quite different plant. It was not found again until 1917, then just inside the Yugoslavian border. It had to wait until 1929 to be introduced to cultivation by Dr Giuseppi who found it in the Nidje Planina. Despite its restricted distribution and its distinct appearance, *C. formanekiana* is not temperamental. Placed in a reasonably sheltered position, it will produce its flowers over a long period. It is a beautiful plant which is easy to grow.

Seed is set plentifully, as is the general rule with biennial campanulas (it is sometimes the third year before it flowers), and if this is collected will germinate well—if it is not collected, replacement plants will be thick around the parent before winter sets in, and these may be thinned out or potted up without difficulty.

We have also had the experience which has been documented elsewhere of having a young rosette of this plant quite eaten away overnight by a snail or slug, leaving but a barren stump; this has later produced a mass of new branches like a pollarded tree, and the plant has subsequently flowered for three years running. We do not, however, recommend this as a standard system of cultivation!

Campanula formanekiana was given an AM in 1931, and is well described and depicted in the *Botanical Magazine* at t. 9436 (1936).

SYNONYM: *C. cinerea* Formanek.

C. FRAGILIS Cyr. 10 cm × 15 cm Pale blue June

'The unbranched stems are heaped and piled with beautiful ample open starry cups of blue',[3] says Farrer. He also says it is far too rarely seen. The reason is that it is hard to place. It is, in fact, probably best in a hanging basket. It can also be good in a warm wall or on the bench of an alpine house. The problems of cultivation are not so much hardiness, but the length of the easily broken stems. *Fragilis* really does mean 'fragile' rather than tender. Under glass it requires little attention, and it repays with pale charming flowers and an elegant shape.

The rootstock is woody, and the stems arise from the base. They have shiny, almost fleshy, dark green leaves, rather like a celandine. They are rounded heart-shaped, approximately 1 cm, and regularly notched. The lower leaves have very long petioles, the upper ones less so.

Like its near relative, *C. isophylla*, the flowers are a large, 2 cm open bell, with a prominent style which extends beyond the petals. These are a delicate light blue with a lighter centre. The narrowly pointed calyx segments are almost as long as the petals. This is one distinguishing point between it and *C. isophylla*, the other being the glossy hairless leaves, which are also generally smaller and of darker colour than its relative. This said, there is also in circulation a plant described as *C. fragilis* 'Hirsuta' which, as the name suggests, is covered in all its parts, leaves and stems with copious silvery bristles and which, incidentally, must be propagated vegetatively, as the seed it sets are virtually always productive of the hairless type. *C. fragilis* has received an FCC. It grows on limestone rocks in Central and Southern Italy.

Division of the woody root is difficult, and so spring cuttings or seed are the best means of propagation.

SYNONYM: *C. barrelieri* Prest.

C. GLOMERATA L. to 75 cm × 30 cm Purple, white June/Sept

This is a popular and readily available border plant, variable but sturdily perennial; it is one of the few campanulas indigenous to Great Britain, and is found throughout Europe and Asia.

C. glomerata is clump-forming, and the basal leaves, usually longish, heart-shaped on stalks, and frequently disappearing at flowering time, throw up stiff stems between 10 and 75 cm high. The stem leaves are narrower, shorter stalked and even clasping the stem, especially near the apex. They bear clusters of flowers in the axils, and the stem terminates with a similar dense cluster. The flowers, which tend to lose their individuality in the cluster ('glomerata'), are bell-shaped, upright, and up to 45 mm long. The calyx lobes are lance-shaped half the length of the petals and without any appendages. The three part style does not extend beyond the petals.

The rootstock, creeping naturally but not dangerously so, except in some light soils, is easily divided. Margery Fish calls it by its country name 'Peel of Bells' and says 'it would, I think, be elevated to a most distinguished position in the plant world if it did not run so badly . . . yet when it flowers I cannot help thinking what a very handsome plant it is. I do not known many flowers in that rich deep blue, and the clustered heads, on stout stems are most welcome in early summer . . . I also like the white flowered form var. alba. White flowers, especially those that do not need staking, are always most useful in the garden.'[4]

14 Campanula glomerata 'Alba' at Bressingham Gardens, Norfolk

The type plant is excessively leafy, but selected cultivars, as follows, are very attractive, and make good border plants.

C. g. var. *alba*
White, single flowers, offered also under various but not always distinctive names, such as:

C. g. var. *a.* 'Schneekrone' ('Crown of Snow')
50 cm. A good white form originating in Germany.

C. g. var. *a.* 'Snow Flake'.

C. g. var. *a.* 'Snow Queen'.

C. g. var. *dahurica*
From south-east Siberia, and one of the best. Tall, 75 cm, flower heads of deep purple (AM 1965).

C. g. 'Joan Elliott'
Violet; dwarf, and early flowering.

C. g. 'Purple Pixie'
Violet-purple; dwarf, with good stem clusters; late flowering.

C. g. 'Superba'
Violet-purple, and one of the best coloured; tall, 60 cm, beautiful, but one of the more invasive (AM 1954).

C. g. 'Acaulis' and *C. g.* 'Nana' are other variable dwarf forms, which frequently come true from seed also, some are good; there are 'alba' variations. A wild pink form has been noted on Salisbury plain.[5]
C. g. 'White Barn' and *C. g.* 'Wisley Supreme' are offered by one or two nurseries. Double forms have been described, but as the flower heads are already clustered any effect is largely lost.

Most may be cut back after flowering, when they will repeat, if less strongly.

SYNONYMS: *C. eo-cervicaria* Nab.; *C. maleevii* Fedorov; *C. aggregata* Willd.; *C. cephalotes* Fisch.

C. GROSSEKII Heuffel. 80 cm × 30 cm Violet July/Sept
This plant ran into trouble when Farrer got hold of its name. Of his two great loves—words and plants—the former won the conflict and the latter gained its reputation—as gross. Grossek was in fact an amateur botanist and, like the plant, of Hungarian origin.

From an evergreen rosette of coarse, heart-shaped leaves on long stems, rise bristly stems bearing similar but stemless leaves. The stem is branched, and bears short-stemmed violet-blue funnel-shaped flowers, either solitary or in small clusters. They are about 3 cm long, and have a hairy calyx which has pointed lobes, with slightly shorter, curved and pointed appendages. The pollen is characteristically bright yellow.

Like *C. trachelium*, which it resembles, it is a plant of stony woodland places, and so looks at its best in the wild or woodland garden, and certainly in a well-drained and poorer soil.

Seed is plentifully set, and this is the best source of the plant in the first place, and of its subsequent increase. Were the plant more widely cultivated no doubt white versions would crop up, and even perhaps doubles, which would make it more interesting. It can be up to 1 m (3 ft) in height in cultivation, and will take quite heavy shade.

15 *Campanula grossekii*

C. INCURVA Aucher 50 cm × 76 cm Lilac, Pale blue July/Sept

C. incurva comes from Greece and certain Aegean islands, and, although biennial or monocarpic, is both easy and hardy, and is well worth growing. The basal rosette, from a stout taproot, is of very characteristic pale green with long-stalked, heart-shaped to triangular leaves. A number of hairy, slightly branching stems curve out and upward ('incurva') bearing gradually smaller leaves which become stalkless towards the apex. The flowers are very pale lilac, lavender or pale blue in lax panicles each 5 cm long and bulbously bell-shaped. The calyx has spreading triangular lobes and short roundish appendages. The three part style does not project beyond the petals.

Seed is flat and brown. It is set copiously in the garden, and forms, of course, the only means of propagation. We have found germination rates to be good and the plant self-sows copiously where it is at its happiest; that is in circumstances close to its native haunts in the Greek sun and on hard limestone. On the wet clay of an unimproved border it coarsens to show disapproval, but it will still flower, and we have seen it through cold wet winters of frosty (down to 16°C) weather. A slightly raised bed with plenty of grit and some lime will show *C. incurva* at its best.

An *alba* form is described, but as the type is in any case very pale, this is not particularly distinguished. The finest form in our experience is pale lilac with slightly deeper lilac edgings to the petals; but with seedlings this is pure pot-luck. Seed is quite frequently offered in the exchanges of the specialist societies (see p. 138). We think this plant well worthwhile; so evidently did the Royal Botanic Gardens, Kew, when they exhibited it in 1937 and it gained an AM.

SYNONYM: *C. leutweinii* Heldreich.

C. ISOPHYLLA Moretti 15 cm × 20 cm Violet, White June/Sept

This is the easy-going plant of cottage window-sills and hanging baskets. Even neglect does not disturb its long and brilliant flowering in late summer.

The rootstock is woody, but the stems are soft and carry broadly oval almost rounded leaves with a heart-shaped base. They are mid-green, long-stalked and with toothed edges. All the plant's leaves are roughly the same shape and size; they do not get narrower higher up the stem as in many campanulas—hence the specific name—isophylla. Most forms are hairless, but some are very slightly hairy. The large round starry flowers are held erect in loose clusters. The calyx tube is short and so the three part style protrudes dramatically. The calyx itself is about half the length of the corolla, triangular and pointed.

This trouble-free plant can be propagated by spring cuttings or from seed. It will often survive outside; Margery Fish grew it outside successfully in

Somerset, but as it comes from Central and Southern Italy it dislikes too much winter damp, and cold winds can break the soft stems.

C. i. 'Alba'

This is a lovely white form, which received an FCC in 1888. We have seen this grown in an oil can in a remote Turkish village shaded from the sun by a tarpaulin. Fred Streeter, the much loved BBC Gardeners' Question Time personality of the 1950s, recalled it. 'I well remember having to get up in the night to water this plant because my mother was afraid the campanula would be dry before morning, and would suffer. Oh, I didn't mind—but it made me careful to see it did not happen again. That's the way to teach you gardening!'[6]

16 *Campanula isophylla* 'Alba'

C. i. 'Balchiniana' (erroneously *C. i.* 'Mayi')

Frequently cultivated in its variegated form, this plant is a hybrid between *C. isophylla* 'Alba' (pollen parent) and *C. fragilis* (seed parent). A plant was given to Messrs Balchin, who took cuttings, one of which threw a variegated shoot, and this form gained an AM in 1896.[7] It is softly hairy, with broadly oval to heart-shaped leaves. The youngest leaves are often pink-tipped, and the cream variegation comes and goes. The plant can sometimes appear to have lost its variegation completely, but then new shoots will be produced with a cream edge. There is a form in cultivation with a broad cream variegation which appears more stable. The flowers are pale blue.

C. i. 'Mayi'

Shown by Mr H. B. May of Edmonton, London in 1899.[8] This was a good form of *C. isophylla* with showy heliotrope-grey flowers, darker towards the tips. Contemporary descriptions do not mention any variegation on the foliage. It received an AM in 1899 but is probably no longer in cultivation. The name has now become erroneously applied to *C. i.* 'Balchiniana'.

C. KEMULARIAE Fomin 30 cm Blue July

This plant, endemic to the Caucasus, in cultivation falls between the border and the rock-garden; but, because it can make a neat front row to the former, especially if tumbling forward over a paved edge, it is included here.

It is allied to *C. raddeana*; a perennial with a thick creeping rhizome. The basal cluster of leaves are oval with a heart-shaped base. They are sharp pointed, doubly toothed, and on stalks longer than the blade. The stems are some 30 cm long, and much branched, bearing oval leaves incised at the base and pointed at the tip on diminishing stalks. The flowers are on long pedicels, of a sometimes rather dullish blue colour—this seems to depend on the type of soil in which they are grown. The natural habitat is on limestone in open forest, and this may give a clue to producing a better plant. The corolla is bell-shaped, with a style which extends beyond the petals and bears orange pollen. The calyx teeth are smooth, triangular, and with no appendages.

C. kemulariae will bear its flowers more proudly if grown on soil which is not too rich. It is also recommended for growing on a wall.

C. LACTIFLORA Bieb. to 150 cm × 60 cm Blue, Pink, White July/Sept

When Farrer, Crook and Graham Stuart Thomas use terms like 'superb', 'one of the finest . . .' and 'one of the best dozen border plants', one stops to ponder . . .

The rootstock is branched and fleshy. The oval leaves, usually mid-green, are thin in texture, with little if any petiole. The strong erect stems have abundant oblong leaves, and branch to produce a broad leafy panicle of erect, individually modest-sized flowers, 2½ cm long, but in sufficient numbers to form a striking head. The individual flower is widely bell-shaped, milky blue ('chain-store, pasteurised Friesian')[9] shading to a white centre. The flaring petals of the bell are divided to half-way down, and have a short style within them. The calyx lobes are largish, but have no appendages.

This plant will do well in the open sunny border, but really prefers partial shade, and must have a moisture-retentive soil to perform well. Because of its profuse seeding (E. A. Bowles spoke of its becoming a weed at Myddelton House), it should be dead-headed. Bowles also recommended cutting it back to just below the lowest flowers. This can increase the flowering period for a total of up to 10 or 12 weeks in all, which is, by any measure, good value for money. The plant does not usually need staking, despite its height.

As indicated, seed is set, but is not easily saved, as the bearing, habit and habits of the plant ensure that all seed is scattered just one hour before attempted collection! Self-sown seedlings may be moved with some success whilst very small, but this plant does not like being moved without sulking for a while, and the eventual plants so obtained are never quite as good as the original. If potted up as a seedling, we have found it best over-potted until planting out, to give minimum disturbance to the root system. Named varieties may be increased by cuttings taken from new growth in spring. It is difficult to divide the thick tenacious roots, and the plants object to it. The type was awarded an AGM in 1926. Robin Lane Fox had a tip for seed collection, 'Oddly, it sets ripe seed within a month of flowering and can be raised by the 100 if you sow it quickly in the same month. Last year I put the admirably easy Lily 'Enchantment' amongst and in front of it. Its fiery orange red and the Campanula's milk blue were all the better for each other. Pale bluish-white hollyhocks rounded off this easy and happy accident.'[10]

C. l. 'Alba'

A white variant; not often found as a true white, as opposed to light grey.

C. l. spp. *celtidifolia*

From Siberia, has smaller but bluer flowers.

C. l. 'Coerulea'

A deeper, China blue (AM 1901).

C. l. 'Loddon Anna'

A fine flesh-pink sport raised in Carlile's nursery. It does not self-seed (AM 1952).

C. l. 'Pouffe'

A sport from Bressingham; mid-blue, long flowering, and less than 30 cm tall (AM 1967).

17 *Campanula lactiflora* 'Alba' at Bressingham Gardens, Norfolk

C. l. 'Prichard's Variety'

Very choice. Dark purple-blue bells, 75 cm. From Prichard's nursery (AM 1964).

C. l. 'White Pouffe'

Another aptly named hummock from Bressingham, a white sport of *C. l.* 'Pouffe'.

OTHERS RECORDED: *C. l.* 'Superba'. Large violet-blue flowers (AM 1969).

SYNONYMS: *C. biserrata* C. Koch; *C. celtidifolia* Boiss. & Huet.

C. LANATA Friv. 70 cm Cream July/Aug

This really is as woolly as the name suggests; a fine, short-piled grey flannel on large heart-shaped, pointed leaves on long, equally hairy petioles. The stem is branched from the base, giving the effect of a tall central spire with lesser spokes lying on the ground. The prostrate stems often turn upwards at the tips. The stem leaves are much smaller than those at the base, and on short petioles. The upper leaves are quite stalkless. In the leaf-axils, and on short pedicels, are held good-sized 'Canterbury Bells' of palest cream, or sometimes palest pink, shades which contrast well with the grey flannel of the leaves. The corolla lobes are bearded with long hairs inside. The calyx is bulky, leafy, with triangular lobes, and long pointed appendages, which turn back on themselves. The three part style is shorter than the petals, and is sprinkled with a bright yellow pollen which serves to help recognise the plant.

This is a handsome plant, but has the misfortune of being either biennial or monocarpic. Seed is, however, abundantly set, easily collected and will germinate well in the following spring if sown in late winter. Its woolliness does not give the sensitivity to winter wet that would perhaps be expected, but a position protected from the extremes of sogginess which could otherwise afflict it will provide better plants and a more satisfied grower. *C. lanata* comes in nature from rocky clefts and cliffs of Bulgaria and northern Greece, and also Yugoslavia. Being a high mountain plant it is hardy to considerable extremes of cold, so long as its roots are protected from too-often repeated freezing and thawing of a wet soil. In Cambridgeshire clay it does well in an open border which is reasonably well-drained. Seed is, of course, the only way of acquiring and propagating *C. lanata* and this is usually offered by the specialist societies which we list on p. 138; and one or two specialist growers, whose names are to be found in *The Plant Finder*, published by The Hardy Plant Society, generally offer it.

SYNONYM: *C. velutina* Vel.

C. LATIFOLIA L. to 100cm × 50 cm Blue, White June/Sept

A dense mass of fleshy roots form a clump of oval to oblong leaves with a heart-shaped base. They are toothed, have a pointed tip and are on stalks of varying lengths. Firm, straight, upright and unbranched stems bear similar oval, stalkless leaves. Both leaf and stem are slightly hairy. Borne in the leaf-axils, and also in a terminal cluster, are the longish bell-shaped flowers, up to 6 cm in length, and held either upright or pendent. In the type they are either light blue or pale lavender in colour. The calyx lobes are about half the length of the corolla and are without appendages.

This is a native of most of Europe, including the northern part of Britain and central Asia; it is absent from the Mediterranean area. It often inhabits woodland or lush meadows, and can be invasive. The cultivars, selected for colour, have the virtue of being less rampant. They like a moist, rich soil, and do better with some shade, though they will also grow well in full sun. *C. latifolia* makes a good border plant and, while the flowering period is not long, it makes an attractive show. As David Stuart and James Sutherland said in their book, *Plants from the Past*, 'its toughness in the face of competition and difficult conditions make it a most elegant plant for wild or woodland gardens. We found the white one among the nettles when we took over our 17th Century walled garden.'[11]

Propagation may be by division, preferably in spring; seed is plentifully set, and resultant seedlings will, of course, vary. We have found that *C. l.* 'Brantwood' seeds true to colour, and also divides very willingly.

The following cultivars are well known:

C. l. 'Alba'
Single white; height variable, does not come true from seed.

C. l. 'Brantwood'
One of the best forms; deeper blue, 70 cm. Longer flowering and will repeat if dead-headed.

C. l. 'Eriocarpa'
Said to be of Russian origin. Purple, 50 cm. Less leafy, neat, and slow to spread.

C. l. 'Gloaming'
A pleasing smokey blue, 60 cm. Not rampant.

C. l. 'Macrantha'
Large, deep blue flowers. Up to 100 cm. Stem sparsely furnished with leafage because of the long internodes, hence tending to look bare and gaunt. Possibly a tetraploid.

C. l. 'White Ladies'
A good white, with large flowers; a longer flowering period than some. 70 cm, and well-behaved, like a perfect lady.

Others
C. l. 'Macrantha Alba' has been described in catalogues as a white 'Macrantha'.
C. l. 'Pallida'—a very pale blue.
C. l. 'Schneeglocke' ('Snowbell')—probably the same as 'White Ladies'.

SYNONYM: *C. eriocarpa* Bieb.

C. LATILOBA A. DC. 90 cm × 30–45 cm Blue, White July/Aug
This is a plant which is very closely related to *C. persicifolia*, but to the gardener there are significant differences. *C. latiloba* is more robust and coarser, enabling

18 *Campanula latiloba* 'Alba'

it to display a strong sweep of colour in a large border—not a role for which *C. persicifolia* is very well suited. It is distinguished by its evergreen rosettes, individually similar to *C. persicifolia*, but larger and formed in clumps. A further distinction is the quite stalkless flowers of *C. latiloba* which are held on stiff rather thick angular stems, 10–15 mm in diameter.

The long narrow leaves of the basal rosette are lance-shaped and toothed, with slightly winged petioles. They can reach up to 25 cm long, 16 cm being more average. As usual stem leaves are shorter and clasp the stem; they are hairless. Calyx lobes are broadly lance-shaped to triangular and hairless; they have no appendages. Flowers are 3–4 cm and widely bell-shaped, quite stalkless and deep blue. The seed capsule opens by three pores in the middle.

A patch of this plant forms a good ground cover, and whilst only the stronger rosettes form flowering stems, these do not usually need staking, and if blown over, they tend to kink and turn upwards once more. Flowering is over a fairly long period, and the plants will repeat if cut down by half after the first flush. On the debit side, the stiff stems and stalkless flowers make the plant something of a clumsy country bumpkin compared with its elegant cousin *C. persicifolia.* Margery Fish said, 'As a child this was the only campanula I knew and it was the whole family of campanulas to me. For years we hardly saw it except in tiny gardens, but it has once again come into its own in a better, deeper form, as *C.* Highcliffe.' (sic)[12] and 'I would call *C. latiloba* a ramper, it certainly is with me . . . I am always finding stretches of its tufted green leaves in places where I certainly never planted it. It will send up its 18 in (46 cm) flower spikes in the most inauspicious places, the arid soil under hollies or poplars, the dry soil under walls, and I am sure would have a shot at covering any piece of waste soil. In these cases I find no difference in the energy of the blue and white forms.'[13]

Flora Europaea makes this plant *C. persicifolia* spp. *sessiliflora*. It is a meal of a name, and they have not convinced us. To the gardener there are such distinctive qualities that this diagnosis would never have occurred. We note, incidentally, that this plant is not a native of western Europe, but of the Balkan peninsula, Turkey and the Caucasus. The *Flora of Turkey* gives it as *C. latiloba*, while the Russians, who tend in general to be 'splitters', make it a subspecies of *C. persicifolia.*

The following are cultivated:

C. l. 'Alba'

A particularly brilliant white, with a slim and more delicate shape. It is rare but worth the search.

C. l. 'Hidcote Amethyst'

Received an AM in 1965 when shown by Alan Bloom. A lilac-pink mutation that was named after Hidcote Garden, in Gloucestershire, where it is still grown and at nearby Kiftsgate. The colour is a pale amethyst shading to a deeper stripe at the centre of the petals and at their tips. Less coarse than the type and very desirable.

C. l. 'Highcliffe'
A rich lavender, selected by Prichards, and a very strong grower (AM 1935).

C. l. 'Percy Piper'
Dwarfer than the type; again, a rich lavender, this has been reported to be a hybrid with *C. persicifolia* but we can find no evidence of the claim; but see above.

SYNONYMS: *C. persicifolia* spp. *sessiliflora* (C. Koch) Velen; *C. sessiliflora* C. Koch; *C. grandis* Fisch. & Meyer.

C. MEDIUM L. 75–90 cm × 30 cm Lilac, Blue, White July/Aug
The well-known Canterbury Bells. A summer border of these large sturdy and free-flowering biennials is a dramatic sight, particularly as an adjunct to a rose garden.

In the first year plants form a rosette of elliptical leaves. They are roughly hairy, with rounded notches, and generally without stalks. In the second year the stout taproot produces an erect stem, much branched, up to 90 cm, with lance-shaped upper leaves. Flowers are lilac or white in wild plants, but in cultivation there are many shades of blue through to mauve, purple or rose. They are showy, more or less erect, and wide at the base. The bells are about 6 cm, oblong with stubby flaring lobes. The calyx has large heart-shaped and reflexed appendages.

C. medium is largely a mountain plant from Southern and Eastern Europe, but is naturalised elsewhere. It is, for instance, established on railway embankments in the South East and East Midlands of Britain.

Sow in April in a nursery bed, and sift silver sand over the seed. Prick out into rich soil, 20 cm apart. Lift and plant out in the autumn. Alternatively pot up and grow in a conservatory where it is very effective for late spring.

In cultivation there are two different double forms, where the calyx becomes petal-like. In one this forms a double bell, one inside the other, called cup-in-cup or hose-in-hose. In the other the calyx is spreading, and forms a saucer to the bell's cup.

Seed is available from Holland of a single-flowered strain of blue, dark rose, lilac, pink and white. Double-flowered strains are also offered in blue, pink and white; and in cup-and-saucer forms in the same colours and also lilac. Specialist seed firms in Britain offer most of the above, and also a dwarf strain. It is a matter of individual choice. Robin Lane Fox said 'My family do not

19 *Campanula medium*

share my love of the double forms, especially in shades of ink purple, but you may agree with me in liking their gross vulgarity'.[14] In the eighteenth century there were striped blue and white Canterbury Bells, but these seem to have been lost.

C. medium has received many Awards of Merit over the years. *C. m.* 'Calycanthema' (the cup-and-saucer form) was exhibited by Veitch and received its AM in 1889. In the same year they exhibited *C. m.* 'Flore Pleno', presumably a cup-in-cup which also received an AM. There followed *C. m.* 'Flore Pleno Roucarmine' 1929, *C. m.* 'Mauve' 1929, *C. m.* 'Meteor' 1915, *C. m.* 'Single White Improved' 1971 and *C. m.* 'White' 1929 from Webb and Dobbie. We must assume that they were all seed strains.

This easy biennial has rather fallen from favour. It is perhaps rather hairy and coarse, and its dead flowers hang on the plant, so that it benefits greatly from dead-heading. Its advantages, however, a mid-summer flowering (so that it does not leave the gap caused by removing spent wallflowers or sweet Williams) and the lovely soft shades which can be relied upon to produce a sea of colour whatever the weather. A large border could be interplanted with a *Dahlia* like 'Coltness Gem' which would give a continuation of display.

SYNONYM: *C. speciosa* Pourr.

C. MIRABILIS Alb. 46 cm × 22 cm Lilac, Blue July/Aug

The student of etymology will recognise 'marvellous' in the specific name, and the plant, so little known today, is more than worthy of it. *The Flora of the USSR* says: 'A rare, very beautiful plant, cultivated since 1898.'

The Russian botanist Nicolas Albov found just one plant of this campanula in the Caucasus; he brought it to the Boissier herbarium in Geneva for preservation; Henry Correvon found among the dried material two carpels with ripe seed in; this gave the first cultivated plants, a sample of which, shown in 1898, immediately won a First Class Certificate.

The large flat basal rosette is made up of smooth, glossy, dark green leaves, oblong (widest at the apex), with a long flanged petiole. The margins have rounded notches and are furnished with stiff transparent hairs which look like prickles—a feature which makes this campanula instantly recognisable even when quite tiny. Eventually a stiff pyramidal central stem develops with a few branches from the base, and these are clothed with alternate leaves similar to those of the rosette but wider at the base and stalkless. Short branches arise from the leaf axils bearing up to four large pale lilac flowers occasionally as much as 10 cm across, but usually 7 cm and of a good bell shape. The petal lobes are one-third the tube length, and the style does not project. Calyx lobes are broadly lance-shaped or triangular, the appendages are also triangular and as long as the calyx tube. The edges of the calyx lobes bear translucent prickles like those of the leaves. Albov's original plant bore more that 100 flowers.

Veitch wrote: 'This plant (in a pot) is grown in pebbles with very little soil, and this is the only way we have been able to get it to flower, and we think this is the secret of success with this kind. The seed was sown three years ago.'[15] Crook is of the opinion, however, that it will not give of its best so grown and needs a richer soil in a position, such as a rock crevice, where the neck will be protected from damp. *C. mirabilis* is a lime and sun-lover. It is very suitable for pot cultivation where, well treated, it will make a grand show.

Propagation must be from seed, which is plentifully set. It may take 3–4 years to flower when, being monocarpic, it dies.

C. mirabilis is from Transcaucasia and has been described as a tertiary relict. Occurring as it does at the epicentre of the genus, closer study of this species, especially from an evolutionary point of view, might well produce interesting information on the genus as a whole.

C. PATULA L. 30 cm × 20 cm Violet-blue, White July

C. patula is one of only five British native campanulas, which is a good reason for its inclusion in this book. Another good reason is Farrer's description in *The English Rock Garden* of it 'filling the alpine meadows with a tossing sea of hot lilac-lavender'.[16]

This is a plant whose leaves and stems are scabrid (thin and dry), but they are very many when well grown in the garden, and the effect is of a profuse and starry harebell. It is generally accepted to be a biennial; although *Flora of the British Isles* says it can be perennial, this has not been our experience.

The lower leaves are more or less oblong, wider at the tip, their edges notched. The base of the leaf becomes a short stalk. Stem leaves are smaller, narrower and stalkless, their margins less notched. The flowering stems are branched and spreading, their flowers often very numerous and held erect. The lobes of the corolla ($1\frac{1}{2}$–2 cm) are about as long as the tube, which narrows at the base. The calyx teeth are triangular and held erect. Differences in the margins of the calyx teeth, more of interest to the botanist than the gardener, give the subspecies mentioned below. When the seed capsule is produced it opens with pores at the top. Again, Farrer's description gives the best impression 'Its stems can attain to 2 or 3 ft [60–90 cm], wirily slight and slender in growth, set all up with oval-toothed leafage, veiny and flimsyish; then crowning the stems in a loose shower are several, or many, wide, erect, and particularly full-rayed stars, or shallow bells, of a luminous rose-purple, varying to palest tones and a stainless white.'[16]

C. patula is a plant of shady woods and hedge-banks, and is scattered throughout Britain. It also occurs in most of central and northern Europe, but not in Iceland.

20 *Campanula patula*

C. patula ssp. *abietina* var. *vajdae*
Sometimes found referred to as *C. vajdae*, is a dwarf mountain form of *C. p.* ssp. *abietina*.

Flora Europea gives the following subspecies:
C. p. ssp. *abietina* (Griseb.) Simonkai
C. p. ssp. *costae* (Willk.) Fedorov
C. p. ssp. *epigaea* (Janka) Hayek
C. p. ssp. *patula*

C. PERSICIFOLIA L. 90 cm × 30 cm Lilac-blue July/Aug
The peach-leaved bellflower is easily brushed past unnoticed in its native alpine woods, but in the garden it is altogether larger and more bountiful. It is a refined and good mannered plant and would probably be more treasured if it were rarer.

On the ground there is an evergreen rosette of hairless linear to lance-shaped leaves with slightly notched margins. Underground it forms a white fibrous creeping root system. The wiry flower stems have a few smaller leaves, and from their axils the flowers appear on short pedicels. They open at the top first, in an elegant profusion of chubby bells. The flowers are 2–4 cm across, with wide-open cup-shaped bells that are slightly nodding.

A native of Europe, Africa, North and West Asia, it was introduced in Britain in the sixteenth century. It is established in South Devon, Gloucestershire, Berkshire and possibly Yorkshire and Grampian on commons and open woodlands.

Cultivation is quite easy, but not invariably reliable. The plants can be mauled by rough weather when flowering and are not easy to stake. Some of the cultivated forms are susceptible to rust attack in late summer. Regular spraying as soon as any rust is noticed is necessary. It is easily recognised as bright orange pustules the size of a pin head on the under sides of the leaves. Triforine (ICI Nimrod T) is effective. A moist rich soil over chalk is ideal but not essential. Like *C. pyramidalis* it can be grown in pots plunged in frames over winter, to be brought into a greenhouse for spring flowering. The more delicate cultivars are probably worth giving this treatment. Propagation by division is easy—it spreads, but does not really run, and seed germinates freely. It is a good cut flower.

C. p. 'Alba'

A white form of the above. Different colour shades will be produced in any batch of seedlings, though most will be lilac-blue.

C. p. 'Carillon'

A tall version of *C. p.* 'Alba', 106 cm × 30 cm and introduced by Kelways. Flowers 6 cm wide × 4½ cm deep. Leaves 15 cm long × 1½ cm wide and deeply rounded notched. Resistant to rust.

C. p. 'Grandiflora Alba'

The plant is large, 122 cm × 60 cm, with big flowers with more pointed corolla lobes than the type. A suspicion of purple gives softness to the white. It looks well from a distance, or grown as a specimen, since it is very floriferous. Grown in one or two gardens (AM 1890). Possibly the same as *C. p.* 'Backhousei'.

C. p. 'Snowdrift'

Single white. Recommended by Frances Perry and still grown in some gardens, and recently available from Blooms.

C. p.'Telham Beauty'

This is a classic herbaceous border plant, which forms a sturdy block of colour for many weeks. It is 90 × 60 cm and a vivid lilac-blue. A thicket of ample spikes carry 7 cm-wide shallow lavender cups, in healthy profusion. Whereas

C. persicifolia has 16 chromosomes, *C. p.* 'Telham Beauty' has 32, and this larger size seems to relate to this doubling.[17] Its history is well documented. In 1705 something very like it was illustrated in the *Botanical Magazine*, as var. *maxima* from South Carolina. Then it vanished. During the First World War, the gardener at Telham Court, Battle, in Sussex, Mr F. D. Thurston, reproduced the plant. His employer Captain Lambert was killed in action in 1915. A year later the plant 'Telham Beauty' received an AM. That plant was reported as having 10 cm flowers, 12 to a spike, and was grown in pots for the conservatory. Soon it was grown by the thousand for the cut flower market near London. Its raiser recommended a light rich soil, division after flowering and a fungicide spray against rust. He also bred a pure white form. But in 1920 he moved jobs to Horsley Hall Garden near Wrexham, and the white plants were stuck for 14 days in a furniture van and died. Possibly the plant we now grow is not Mr Thurston's original which may no longer exist. A lot of inferior plants masquerade under its name. It is self-fertile.

C. p. 'Donald Thurston'

This form was bred from *C. p.* 'Telham Beauty' and received an AM in 1926 when exhibited by Mr Thurston. It had very large pale blue flowers, but sadly does not appear to be in cultivation now.

C. p. 'Wedgwood'

This was raised from seed and exhibited by Watkins and Simpson in June 1953 when it received an AM. The stiff erect stems are 112 cm high and hold up to 50 flowers per stem, with 10–12 open at once. The flowers are 7 cm across and 2½ cm deep, a violet-blue with conspicuous cream stamens. A plant named 'Wedgwood' is available from a few nurseries, but whether it is the same as the original cultivar is in some doubt.

C. p. 'White Queen'

Another strong white form which is available from several nurseries.

C. persicifolia var. *planiflora* 15 cm × 30 cm Lilac-blue or White July/Aug

This is a rather smart little plant, but it just misses being first class. It is rather too stiff, and the flowers are disproportionately large.

The dark green leaves are thick and glossy, and have notched, slightly overlapping margins. The flowers are flattish, large, and held close to the stem. A curious feature is that, unlike other campanulas, the ovary appears to be superior, since it rises above the sepals. This is both misleading and disconcerting. The plant's history is also interesting. First illustrated in Dodart's *Histoire des Plantes* in 1676, it was then called 'Trachelium americanum minus flore caeruleo patulo'! It survived that meal of a name, and

1 *Campanula* 'Burghaltii' at Sissinghurst Castle, Kent

2 *Campanula turbinata*

3 *Campanula carpatica* 'Blue Moonlight'

4 *Campanula collina*

Alphonse de Candolle[18] said that it was in many gardens and herbaria, and that the habitat was Arctic America. The Rev. Wolley Dod, of Malpas, challenged this in the *Gardeners' Chronicle* of 1895 'If this alleged species were lost, I would undertake to make it again in a few years by a selection of dwarf forms of *C. persicifolia*'.[19] His practical observations proved right in the 1920s, when the John Innes Institute did some breeding experiments. They showed it to be a Mendelian recessive.[20] Plants of *C. persicifolia* being self-pollinated produce *C. persicifolia planiflora* in proportion of 1 to 8. If *C. p. planiflora* is back crossed with *C. persicifolia* then one *C. p. planiflora* to three *C. persicifolia* are produced. It breeds true from seed. Gardeners, however, clung to the Arctic America story for long thereafter.

It is usually grown in the rockery, but likes a little shade. Blue, pale blue, white, single and double forms have been noted, but the blue and white are now the only ones available. The white form received an AM in 1970 when exhibited by Valerie Finnis. More seedlings appear white than blue in our experience. Available from several nurseries.

SYNONYM: *C. nitida* Dodart.

Campanula persicifolia Double forms

For convenience the double forms are listed together below. The *Botanical Magazine* of 1798 stated that double *C. persicifolia* had become so common as to almost usurp the single ones from gardens. Nowadays rather more elusive, a few nurseries still list them. We give details below of those which we know to still be in cultivation. In the text W = wide, and D = deep.

The 'doubles' can be rather exasperating plants. Sometimes they flower abundantly, but they can sulk too. Most increase quite well in the usual stoloniferous way. Large leaves in the basal rosette usually mean that the plant will flower, but if it does not it is worth lifting in late August, splitting the clumps and growing on in the vegetable garden for replanting in the border the following spring. Liquid feeding and mulching also help, and so does the spray for rust mentioned earlier. They produce a little viable seed, which will produce some double-flowered seedlings, many of which are simply messy looking. It is only worth retaining plants which are strong with a good clear outline.

C. p. 'Alba Coronata' 45 cm × 30 cm White July/August

Rather a dainty plant, whose flowers are usually two, sometimes three, rows of petals. The calyx has become enlarged and coloured white like the corolla. It is in effect a hose-in-hose type flower, but cup-in-cup is more descriptive. Rosette leaves 11 cm × 1½ cm, flowers 5 cm W × 3½ cm D. Reintroduced by Mrs Joan Grout from an old garden in north Nottinghamshire. Sometimes sold as *C. p.* 'Alba Plena'.

21 *Campanula persicifolia* 'Alba Coronata'

C. p. 'Alba Flore Pleno' 60 cm × 30 cm White July/Aug

Very double with many rows of petals. Possibly the same as *C. p.* 'Gardenia' or *C. p.* 'Boule de Neige'. The flowers may decay in wet weather, but as a buttonhole they last very well.

C. p. 'Boule de Neige' 60 cm × 30 cm White July/Aug

A very double flower with outer petals rolling back on themselves a little to give the impression of a ball. Not very strong and a martyr to rust. Rosette leaves 12 cm × 1½ cm, flowers 5½ cm W × 3 cm D. Received an AM in 1921. Available from a few nurseries.

22 *Campanula persicifolia* 'Boule de Neige'

C. p. 'Coronata' 46 cm × 30 cm China blue July/Aug

A semi-double cup-in-cup. The bell is square in section, and shallow, being 5 cm W × 2½ cm D. Calyx lobes linear and flat to the petals. Leaves linear 8 cm.

C. p. 'Flore Pleno' 70 cm × 30 cm Lilac blue July/Aug

A cup-in-cup with two rows of petals, and the centre filled with small petaloid stamens. The effect is of a small button rather like a rose or double primrose. Flowers 4 cm W × 2½ cm D, leaves 13–15 cm × 1 cm. Though the flowers are small they are plentiful. Listed by a few nurseries.

23 *Campanula persicifolia* 'Flore Pleno'

C. p. 'Fleur de Neige' 70 cm × 30 cm White July/Aug

A very large tightly packed flower with approximately 3 rows of petals and petaloid stamens. The corolla lobes are pointed at the tips. It opens rather flat, as its dimensions of 5 cm W × 3 cm D imply. Leaves 8 cm × 1 cm. It is grown in a few nurseries and gardens, but not always under its given name. It is resistant to rust.

C. p. 'Gawen' 45 cm × 30 cm White July

A semi-double cup-and-saucer, pure white and although quite short in stature a vigorous plant. It was rediscovered in a derelict garden in Derbyshire, and is now available from one or two nurseries. Flowers are 4 cm W × 3½ cm D, leaves 11 cm × ½ cm.

24 *Campanula persicifolia* 'Hampstead White'

C. p. 'Hampstead White' 70 cm × 30 cm White, Green July/Aug
A semi-double cup-and-saucer, rather like a distinguished Canterbury Bell. The veins on the backs of the petals and their tips are sometimes tinged with green. Available from several nurseries under different names. Flowers 5 cm W × 3½ cm D, leaves 13 cm × ½ cm. Possibly the same as 'Hetty'.

C. p. 'Hetty' 70 cm × 30 cm White, Green July/Aug
A semi-double cup-and-saucer. A vigorous and floriferous plant which has been re-introduced and named by Dr C. Hardwick in Surrey for his mother. It received a PC when exhibited by him at the RHS in 1985. Flowers 4½ cm W × 4 cm D, leaves 11 cm × ½ cm. Possibly the same as 'Hampstead White'.

25 *Campanula persicifolia* 'Hetty'

C. p. 'Loddon Petal' 60 cm × 30 cm Dark blue July/Aug
A semi-double cup-in-cup. It is not a very strong plant. Introduced by Carliles in the 1970s but not currently available from them. Still apparently grown, but we have not seen it.

C. p. 'Loddon Sarah' 60 cm × 30 cm Lilac-blue July/Aug
A semi-double cup-and-saucer. As above, introduced by Carlile's.

C. p. 'Pride of Exmouth' 60 cm × 30 cm Purple-blue July/Aug
A semi-double cup-in-cup forming a dainty curved bell. This plant has been around for a long time, and therefore must be stronger than its thin stems and small stature suggest. Available from several nurseries, but some inferior plants are given this name in error. It stands up and repeat flowers particularly well. Flowers 3 cm W × 2 cm D.

26 *Campanula persicifolia* 'Pride of Exmouth'

Where are they now?

The following is a list of *C. persicifolia* names which have been mentioned in gardening literature, but which appear to be no longer in cultivation. Most date from the 1920s, but some are earlier.

'Backhousei' S W 1883
'Beechwood' S B
'Blue Bell' S deep B
'Blue Bird' S B
'Coupe d'Azure' D B
'Daisy Hill' B
'Dawn Patrol' D W 1948
'Delft Blue' D B
'Essie' S B 1953
'Everest' S B
'Fairmile' 1949
'Fairy Queen' S B/Grey 1927
'Frances' D W/Lavender AM 1938
'Gardenia' D W 1949
'Humosa' D B
'La Fée' S W
'Lavender Queen' B S
'Marion Knocks' 1955
'Misty Morn' D B
'Moerheimii' D W AM 1900
'Pfitzeri' D B
'Princess Royal' 1935
'Ryburgh Bells' S B AM 1923
'Ruth Lansdell' 1959
'Shirley' D B AM 1925
'Spetchley' S W AM 1921
'The Crescent' S W
'The King' D B AM 1926
'Veneta' D B
'Verdun' D B 1926
'William Larensen' S B AM 1901
'Windsor Bell' D 1950
'Wirral Belle' D B 1949

D = double, S = single, B = blue, W = white, AM = Award of Merit.

SYNONYMS: *C. crystallocalyx* (Adamovic); *C. persicifolia* ssp. *subpyrenaica* from Pyrenees; *C. persicifolia* ssp. *sessilifolia,* syn. *C. latiloba*, and so treated here.

C. PRIMULIFOLIA Brot. 90 cm × 30 cm Blue July/Aug

'This is a good plant which deserves to be better known than it is, for it is eminently handsome and quite hardy, and may be made a great ornament to our flower borders.' So said the *Botanical Magazine* in October 1855, and so, still, say we! When Farrer dismissed it as summarily as he did, we wonder if he knew and grew it.

C. primulifolia comes chiefly from Portugal, principally in the Algarve and Beira, and also from Spain. It grows in sandy, moist soil in the shade of rocks.

The basal rosette looks exactly—but exactly—like that of a vigorous cultivated primrose. Fresh stems arise each year, branching only from the base and so forming a pyramidal shape to the plant. These stems are angular and succulent. Stem leaves are similar to those at the base, but narrow to a much shorter winged, primrose-like stalk, and they become stalkless toward the apex. Flowers are borne terminally and in the leaf-axils in bunches of 3–5 on short pedicels. The lower flowers are noticeably broader than the upper ones, but all are good broad bells, spreading at the mouth to as much as 5 cm across, and are of a handsome purple-blue with a paler base, somewhat reminiscent of the chimney bellflower. Calyx segments are lance-shaped, backward curving and without appendages. The style is short and divides into three.

This, in spite of its Mediterranean origins, is hardy. The principal difficulty is obtaining seed, which is rarely offered. Subsequent plants are relatively slow to establish, and we have lost many by allowing seedlings to dry out. If they are treated like the primulas they so closely resemble—kept moist and shaded—they will thrive. One of the writers successfully raises them on capillary matting, where, unless all are well labelled, they soon get confused with the primroses which like the same treatment. Once planted out, preferably in a deep humus-rich soil with some sand—drainage still needs to be good—they need to be kept moist until well established.

Division of the rootstock is possible, but not easy, and care is needed for success. This must be done in spring as new growth commences. This plant is figured in the *Botanical Magazine* at t. 4879 (as *C. primuliflora*).

SYNONYM: *C. trachelium* Brot. (not *C. trachelium* L. of today).

C. peregrina L.

This is a similar plant, but from the opposite end of the Mediterranean, where it grows in Cyprus, Rhodes, Turkey, Syria and Palestine. It is closely related, and is also found in similar habitats—at streamsides or in damp pine forest. The most obvious difference is that here the flowers are darker at their centres, with lighter lobes, whereas in *C. primulifolia* the 'eye' is lighter in colour. *C. peregrina* is biennial in habit. The *RHS Dictionary* wrongly refers to these as synonymous. It is not known at present to be in cultivation (AM 1903).

SYNONYM: *C. lanuginosa* Lam.

C. PUNCTATA Lam. 30 cm × 40 cm Pink Summer

Whilst in some gardens this species can become a weed, in others it tends to die out without apparent cause. And one could be pardoned for wondering why it was formerly known as *C. nobilis*. However, when successful it does bear beautiful long bells with crimson specks and flecks within. The flowers are waxy cream outside, with the crimson internal spots showing through a little, sometimes to give a rather muddy effect. On a flourishing plant the flowers are freely and regularly hung like the decorations on a sparse Christmas tree.

Soft fleshy pointed heart-shaped leaves on long stalks arise from the rootstock in spring. The stems, 30 cm or more in a vigorous form, bear similar but more or less stalkless leaves, and terminate with the pendent flowers about 5 cm long. The calyx lobes are triangular and lie flat to the corolla. The appendages are oval, reflexed and about half the length of the rest of the calyx.

27 *Campanula punctata* 'Alba'

28 *Campanula punctata* 'Alba': single flower

The inside of the flower is really most attractive, and success with its planting depends on placing it at or above eye-level so that the inside of the bell can be seen. It has a clear preference for a sandy soil or a peaty one, but pines in heavy clay; it seems indifferent to lime.

This campanula comes from Siberia and Japan. Very pale forms have been sometimes selected as 'Alba'.

C. p. var. *hondoensis*

The Japanese form, described as subspecies or var. *hondoensis*, is also occasionally referred to as *C. hondoensis*. It is in general considerably larger than the Siberian form, and the fact that it also lacks appendages may justify taxonomic separation. To the gardener it just makes for a larger, sometimes more ungainly version of *C. punctata*.

C. p. var. *microdonta*

This form is smaller and paler, and the whole plant less hairy.

C. p. 'Rubrifolia'

This form is in circulation; here the veining on the outside of the corolla is also red, while the interior colouring is also more liberally scattered and deeper, so that overall the flower gives a much darker red impression. It is very floriferous.

As may be inferred from all this, *C. punctata* is clearly variable in form. It has also given rise to the better *C.* 'Burghaltii' and *C.* 'Van Houttei'. These are crosses, it is surmised, between *C. punctata* and *C. latifolia*. The former has *C. punctata* as the pollen parent, the latter as the seed parent, it is thought. They are described separately.

SYNONYM: *C. nobilis* Lindley.

C. PYRAMIDALIS L. 122–150 cm × 70 cm Blue, White July

Cultivation of the Chimney Bellflower reached its height in the nineteenth century, and was much grown in Edwardian and Victorian times as a house-plant. Its huge size was its attraction, as it was brought into the hall or drawing room and stood in the empty chimney breast. This idea has rather overshadowed its use as a hardy plant for the back of the border, where it can reach monstrous proportions. It does not please everyone however, Christopher Lloyd said 'The reason they are unsuccessful in the garden is because the bees pollinate their flowers, and these then fade in a matter of three days instead of three weeks. All campanulas are similarly bee-inflicted and are well worth growing as pot plants for indoor use, whenever possible.'[21]

C. pyramidalis comes from northern Italy and the north-west of the Balkan peninsula. It is strictly a perennial, but is short-lived. The plant is glabrous, with usually one, but sometimes several stems rising erect to about 122 cm in the wild. The basal leaves are broadly oval, toothed, on long stems. The stem leaves are stalkless and oval to lance-shaped. The many flowers are small, 5 cm wide, broadly lobed bells. They form a pyramid of numerous flowering spikes. The long flower stems hold the starry flowers erect, the styles thrusting forward beyond the petals. The calyx is smooth and about half as long as the petals.

C. pyramidalis is easy to grow. Seed sown in April and planted out will flower in its second year. In the border it needs a shady moist soil, and adequate staking as it may snap at the base in high winds.

In 1892 a form called *C. p.* 'Compacta' received an AM, as did *C. p.* 'Alba' in 1896. The following notes are from the *Gardeners' Chronicle* of 1845.

'This plant, when properly treated will produce a flowering stem from 8 to 16 feet [250–500 cm] in height, regularly branching from the bottom upwards, and forming a pyramid which when the blossoms are expanded is of singular beauty.

Seed should be sown in March in pans, and then pricked out in rich light soil where they should remain until they begin to grow in the following Spring. The strongest plants may be selected for potting; as the plants are not intended to flower until they are two years old, they should at first be put in pots just large enough to prevent the roots from being cramped, and to induce a slow but healthy growth. They should remain through the Winter in a pot no larger than 8 inch [20 cm]. The treatment during the Winter months is to plunge the pots in sand or ashes in a frame where they can be kept dry. On about 20th March . . . the season has arrived when the plants are to be shifted into pots in which they are to flower, and as they will have to be moved from place to place, the pots should not be larger than one man can conveniently carry when filled with soil.

Let the compost be 1/3 well rotted and dry hotbed dung, put through a coarse sieve, 1/3 turfy loam . . . and 1/3 sand; let these be well mixed together, and have ready a quantity of lime-rubbish, about the size of Filberts, or Walnuts; let the pot be well drained, then over the drainage place a layer of the lime rubbish, then the soil, on which place the plant with the ball entire, and as you proceed to fill the pot let handfuls of the lime-rubbish be scattered round it with the soil . . . great attention should be paid to free the soil of worms. The plants should now be placed in a cold frame, and the light kept close, admitting a little air when the sun shines, with an occasional watering overhead to produce vigorous growth. When the stem has grown 4 or 5 inches [10–13 cm], the plant should be removed to . . . the late Vinery where fires are seldom used. The shade and proper treatment of the Vines in a house of this description are admirably adapted for the vigorous growth and elongation of the flower-stem; when it has attained its greatest height, the plant should gradually be exposed to more light, which will give strength to the stem and colour to the blossom, and if circumstances have been favourable, it will excite the admiration of everybody.'[22]

C. p. 'Aureo-variegata'
A form with yellow blotched variegated leaves is offered by one nursery. This must be propagated vegetatively.

C. RADDEANA Trautvetter 35 cm × 25 cm Deep blue July/Aug
Another Caucasian, bearing the name of its first collector, G. Radde. Running roots form mats of rosettes of dark green shiny leaves which are triangular to heart-shaped, toothed, and on long petioles. A fine stem, often of a deep wine-coloured tinge, arises from each rosette; this carries similar but shorter-

29 *Campanula raddeana*

stalked leaves and a raceme of large, drooping deep violet-purple bells. The petals are lobed to one-third of their length (about 2 cm). The calyx has triangular appendages, and the three part style is the same length as the corolla. The pollen is of a characteristic orange colour.

A thoroughly colourful and attractive small plant which won an AM of Merit back in 1908, when shown by Reuthe, and which should have many more opportunities of revealing its merits today. It is at its best in a well-drained, sunny position, in a soil which is not so rich as to promote leaf growth at the expense of flower; if the front of a small border can provide this, it will thrive. It is seen and appreciated better when nearer to eye-level. It is a strong lime-lover, related to *C. kemulariae*, but without the least tendency to hide its bells amongst the foliage as this can.

Seed is sometimes set, and it will self-sow, but such propagation is not to be counted on; pieces of the rooting runners taken in spring with the beginnings of a rosette attached will not fail to establish quickly.

SYNONYM: *C. brotheri* Somm. & Levier.

C. RAPUNCULOIDES L. 100 cm × 45 cm Blue June/Sept

'The most insatiable and irrepressible of beautiful weeds. If once its tall and arching spires of violet bells prevail on you to admit it to your garden, neither you nor its choice inmates will ever know peace again.'[23] Thus Farrer sums up and dismisses *C. rapunculoides*. Well, ever is a long time . . .

This plant owes its success to a mass of thick white *Dahlia*-like roots which spread at an alarming rate in light soils and penetrate deeply. In addition, it seeds abundantly, and the light seeds are blown a great distance. In 1901 the *Gardeners' Chronicle* advised that 'no thistle or dock . . . is worse, and that no nurseryman should be allowed to sell it. One tradesman who was remonstrated with replied that he had a demand for it—to kill grass.'[24] A recent garden visitor commented on just this ability to kill grass, but we cannot recommend it . . . *C. rapunculoides* roots do in effect quite fill a bed, strangling all competition.

Despite such warning, it is too often a fact that the plant is acquired by accident, usually wrongly named. Then it has to be eradicated. One of the writers did indeed receive it in this way, but it was of a clone that appears to be less pernicious than Farrer's. It took but one season to eliminate it from a bed of $3\frac{1}{2} \times 1$ m ($11\frac{1}{2} \times 3$ ft). This was done with a fork and hoe without resorting to herbicides. However, the bed did look somewhat bare that season! For those who use it, glyphosate (Murphy Tumbleweed) would probably have been quicker.

Having given the warning and the remedy, it has to be admitted that the plant is beautiful.

The basal leaves are long-pointed, heart-shaped on long stalks; these wither when the stems elongate. The stem-leaves are stalkless and broadly lance-

30 *Campanula rapunculoides* 'Alba'

5 *Campanula poscharskyana*

6 *Campanula glomerata* 'Superba'

7 *Campanula latifolia* 'Alba'

8 *Campanula latifolia* 'Brantwood'

9 *Campanula lactiflora* 'Loddon Anna'

10 *Campanula lactiflora*

11 *Campanula latiloba* 'Hidcote Amethyst' at Hidcote, Gloucestershire

12 *Campanula persicifolia* at Hatfield House, Hertfordshire

13 *Campanula rhomboidalis* at Bressingham Gardens, Norfolk

shaped. All leaves have prominent veins on the underside, and are lightly toothed. The flowers 2–2½ cm long, are borne in a raceme, each horizontal or drooping on a short stalk. Calyx lobes are lance-shaped, strongly reflexed at flowering but have no appendages. The style is in three parts and the same length as the corolla, the lobes of which are divided to one-third, pointed and slightly reflexed.

C. rapunculoides occurs naturally in most of Europe, Iran, Caucasus, Central Asia and Siberia. It is found naturalised in Britain, Scandinavia and America, often on waste ground. Occurring over such a range, it is somewhat variable in size, leaf and flower shape and invasiveness!

In Russia it has been considered as a cure for hydrophobia.

C. r. 'Alba'

Not frequently seen; we cultivate this in Cambridgeshire, where it is shorter (45–60 cm), less vigorous than the type and quite manageable. It sets seed, though not freely, and should be propagated by division.

SYNONYMS: *C. cordifolia* Koch.; *C. rhomboidea* Falk.; *C. rhomboidalis* Gorter; *C. trachelioides* Bieb.; *C. lunariaefolia* Reichenb.; *C. setosa* Fisch.

A number of subspecific names have been applied, some of them referable to the above. It may be added that a number of unrepeatable vernacular names have also been used by gardeners!

C. RAPUNCULUS L. 60–90 cm × 30 cm White to lilac April

The rampion (not to be confused with the *Phyteuma* also commonly called rampion) was once grown as an autumn and winter vegetable. The name is a diminutive of the Latin, *rapum*, turnip.

A usually hairless biennial, with a thickened root. The basal leaves are oval, widest at the middle, pointed, 4 cm long, on long winged petioles. The upper leaves are narrower with shorter petioles. The flowers are small and narrowly bell-shaped, 2–2½ cm long, divided by one-third into pointed lobes, and hairless. They are whitish to pale blue, on short thin pedicels and are held in long spikes. The calyx teeth are long (nearly as long as the petals) and very narrow, rather like bristles; there are no appendages. The style is almost as long as the corolla, and it divides into three stigma.

This very variable plant is widespread over most of Europe (except the extreme north), South and Central Russia, Caucasus, Crimea, Turkey, Syria, Iran and North Africa.

The plants flower early and freely, but by the summer they have gone to seed and look rather untidy.

The pungent nutty flavour of the roots has been popular cooked or raw since the Middle Ages. It features in the Fairy Tale by the Brothers Grimm,

31 *Campanula rapunculus*: root

Rapunsel. A pregnant peasant girl yearned for vegetables that she could see in the garden next to her. A high wall surrounded the garden, which belonged to a witch. The young woman became obsessed with desire for the fine salad vegetables, and so she began to pine away. Her husband climbed the wall to steal them, and the first time the witch apparently did not see him. However on the next occasion, she was waiting for him. She caught him and made him promise to bring her his child when it was born, in return for letting him have the salads. The young wife gave birth to a fine girl, who was named Rapunsel after the plants.

In 1918 *The Garden* magazine reported that seed was widely available.[25] It advised sowing thinly in April or May in a rich shady soil. The plants mature in October when the root is about 1 cm thick, and 12 cm long, like a small carrot, but white. It is no longer available as vegetable seed. Seed obtained from Eastern Europe has yet to produce plants for us that seem true to type.

Several subspecies and varieties have been described for this variable plant, including ssp. *lambertiana* Boiss. and var. *spiciformis* Boiss.

C. RHOMBOIDALIS L. to 45 cm × 20 cm Blue July/Sept

This is a neglected plant. It has fallen into a niche where, too big for the small modern rock-garden, too small for the border, and unhappy for long in a pot, it remains untended and so unknown. Above all it is too often confused with *C. rotundifolia*. Unknown that is except by those 'who have seen the alpine fields one undulating sea of sapphire waves beneath its sheaves of noble deep-blue harebells gathered loosely toward the top of those stalwart stems, clothed in rhomboidal toothy foliage all the way up.' So Farrer, whose apt description could hardly be bettered, and who continues: 'In cultivation this beautiful thing is no less stalwart and splendid in any border, but its 2 feet or 18 inches [40–60 cm] fit it chiefly for bolder sweeps in the rock-garden, among such old friends and neighbours as *Anemone alpina*.'[26]

It is firmly perennial, growing from a turnip-like root with branching rhizomes, which spread only slowly. The clump of basal leaves are often persistent in winter, but disappear at flowering. They are stalked, while the stem leaves are stalkless. The leaves, being diamond-shaped, the stems appear better clothed than those of *C. rotundifolia*: they are also much stouter. Flowers are in a raceme, not numerous, and about 2 cm long, divided to one-quarter their length, broadly bell-shaped, and of a deep purple-blue. Calyx lobes are linear, smooth, much reflexed, and without any appendages; the three part style is the same length as the corolla.

By itself it spreads but slowly, but pieces of rhizome with roots may be lifted in spring for increase. Seed is set but, like that of many a beautiful campanula, is not offered except by specialists. It loves the sun and, happy, will flower for a long period. In Cambridgeshire, and in a mild autumn, it has flowered into December.

There were, in Farrer's day, many named cultivars, but these remain untraceable today.

C. ROTUNDIFOLIA L. 40 cm Blue Summer

C. rotundifolia is, to the gardener, an all-embracing term! It occurs in nature throughout the northern hemisphere north of the tropics, and tends to receive a local name wherever it is described. *Flora Europaea* splits it into some 20-odd, to which, no doubt the observant gardener would probably add even more. Some names derive from the area in which it occurs, and whilst we agree with the botanists that differences are real, they are of such a scale that the gardener may wisely ignore them. No doubt if they were more dramatic plants, or something of a challenge to grow, there would be more reason, or excuse, for going into the finer detail of relative leaf shapes and sizes, whether they have erect bells or pendent, and whether these are followed by capsules which may be pendent or erect; permutations the scrutiny of which may well dishearten the gardener while his weeds keep growing . . . There is many another plant which has no special claim to fame apart from the fact that it provides a challenge to the specialist alpine grower; *C. rotundifolia* is not found among such.

One outstanding plant is *C. linifolia* 'Covadonga' (though the linifolia is now

32 *Campanula* 'Covadonga'

obsolete). Collected in what is now the Covadonga National Park in northern Spain, it is a somewhat sprawling little plant with wiry stems and pointed oval basal leaves and longer, finer stem leaves with three or four teeth to each margin, all on long petioles, and terminated by a largish bell of a deep violet blue.

The native British harebell, the Scottish Bluebell—a name which is repeated in most European languages—remains one of our best wildflowers:

With buds, and bells, and stars without a name,
With all the gardener Fancy e'er could feign,
Who breeding flowers, will never breed the same.

Keats

Rotundifolia translates as 'round-leaved'. Many an innocent observer has failed to find the round leaves, which, in fact, are usually withered away at the time of flowering, the remaining visible ones being only variable in how long and narrow they can be. This plant can be grown in any soil, where it will form a large round clump before setting out to spread far and wide, and also seeding quite generously, but no gardener who can operate a hoe need feel menaced. Apart from the varying shades of blue—most frequently pale—there are slim records of doubles, and also white variants; but in spite of the wide distribution of this aggregate these are not often come across in the wild. There are, however, reliable reports in the past of the following:

C. r. 'Alba'

An albino form, which was reported to come largely true from seed.

C. r. 'Flore Pleno'

This is said to have a perfectly double bell, the flowers tending to open very widely, forming as it were a double-edged wall with frilly edge; this was supposed to exist only in blue. The description as given is strangely (or perhaps, not so strangely?) reminiscent of *C.* 'Haylodgensis', which see.

C. r. 'Olympica'

Comes from the Olympics in the state of Washington, USA, where it is said to be very variable, but at its best is a robust 22 cm tall wide-belled campanula with dark green, rather jagged leaves.

C. r. 'Soldanelloides'

With the bell slit into numerous narrow segments, like the corolla of a Soldanella, it was stated to be a 'freak' which the writer (in *The Garden*) had lost.

As this campanula is so very widely distributed, it is hardly surprising that it is so variable; the surprise is that it is not in fact more so. The botanists distinguish many forms, but we tend to finish up where we started this account of the species—all-embracing! William Sutherland, manager of the herb department at Kew, writing in the *Handbook of Hardy Herbs and Alpine Flowers* in 1871 has the last word: 'Though not very "far fetched" this plant so wonderfully increases in beauty under care and culture that I cannot pass it over in this list without strongly recommending it to those who may not have given it a trial.'[27]

C. SARMATICA Ker-Gawl. 50 cm × 40 cm Grey-blue May/July

This species comes from rocky and stony slopes at alpine and sub-alpine levels in the Caucasus.

A large clump of coarse, wrinkled and crumpled, hairy grey-green leaves, triangular and pointed in shape have a heart-shaped base, and petioles that are longer than the leaf-blades. There are several stout unbranched hairy

33 *Campanula sarmatica*

stems, and these have similarly shaped, but smaller, stemless leaves. The pale blue or blue-grey flowers are borne in one-sided conical racemes. They are bell-shaped with flaring lobes, bearded within, each of which shows an even more hairy nerve-line. The calyx is of bristly lobes hugging the corolla tube and half its length. It has short triangular appendages which are reflexed in a continuation of the line of the main calyx lobes. The three part style equals the length of the corolla.

The effect is of a rugged plant, soundly hardy and perennial; perhaps a little coarse, but yet quietly attractive. It is well-placed in the forward part of the border or in the rock-garden, and it has been suggested that it looks particularly well against a background of paler green conifers. It has something of the habit of a smaller *C. alliariifolia*. It can take a little more sun than some campanulas, and is said also to be unattractive to slugs, possibly because of the hairs.

C. sarmatica cannot be propagated by division, growing as it does from a single rootstock; seed is, however, plentifully set, from which fresh plants may be easily raised. It was introduced in 1805, and figured in the *Botanical Magazine* t. 2019.

C. sarmatica should not be confused with *C. sarmentosa*, a synonym of *C. rigidipila* from Ethiopia, which does not now appear to be in cultivation. Sarmatica is an old name for the southern part of European Russia.

SYNONYMS: *C. albiflora* C. Koch (white form); *C. betonicaefolia* Bieb.

C. SIBIRICA L. 50 cm × 30 cm Pale blue June/Aug

This plant occurs over a very much wider area than the name suggests—all over Eastern Europe, through to the Crimea, and over most of Russia. Growing as it does under such varied conditions, it is extremely variable.

It is a biennial or monocarpic species, taking two or even three years in cultivation to flower. It is a hairy plant throughout. A single or several stems, according to the form, arise from a basal rosette of lance-shaped or oblong to lance-shaped leaves with winged petioles. The stem leaves are lance-shaped and stalkless. The stems are sometimes branched above and bear flowers terminally and on short pedicels in the leaf-axils. These flowers may be numerous, or borne singly, and are of a bluish-lilac colour, sometimes nearly white. The corolla forms a more or less funnel-shaped bell 15–20 mm long, with flaring lobes, sometimes hairy within. The calyx lobes are pointed and bristly, the appendages reflexed. It is variable, like many characteristics of the plant, in size and shape. The three part style is about the same length as the corolla.

C. sibirica can, of course, only be grown from seed, but it is not an unattractive plant when grown *en masse*.

34 *Campanula sibirica*

It was introduced by Dr William Pitcairn as long ago as 1783, and grown in his botanic garden at Islington. It was illustrated in the *Botanical Magazine* t. 659.

C. SPECIOSA Pourr. 46 cm × 30 cm Blue, White July
This is a plant which has been neglected of recent years, but which could well fill a useful and colourful place in the narrower borders of today's smaller

gardens. It was introduced in 1820 and much grown for years, and received an Award of Merit in 1928; yet it remains relatively unknown today. Almost anyone could be forgiven for thinking it a dwarf compact Canterbury Bell.

A large, flat rosette of narrowly lance-shaped leaves throws up a stout and bristly leafy stem; the stem leaves are narrower, and stalkless. The relatively large, well-proportioned bell-shaped flowers, 1½–3 cm long, are borne on individual pedicels; these are very long at the base, and gradually decrease in length up the stem, so that the effect is of a pyramidal spike of flowers. The flowers are blue or white. We have not heard of the pink shades seen in the Canterbury Bell, but this shade could well appear, as in other species, if *C. speciosa* was grown more. The calyx lobes are short, narrow with reflexed stubby appendages. The style is in three parts and about the same length as the corolla.

Unlike the Canterbury Bell, which is, of course, biennial, this plant has been described as perennial (*Flora Europaea*) and monocarpic (*RHS Dictionary*) with which Crook agrees. It is certainly best treated as the latter. Although it may take several seasons to flower, like *C. pyramidalis* it does not flower so effectively a second time. Unlike *C. pyramidalis,* it prefers a poorer soil, and it must have a warm sunny spot to give of its best, as it does in the rocky limestone soils of the Cévennes, Corbières and Eastern Pyrenees from which it comes, and where summer holidaymakers may well have spotted it.

If it is to be used in the border, it is best raised in pots with some winter protection to prevent rotting at the crown, though it needs no heat. It can then be planted out in spring, when it will make a magnificent display . . . Grown in a gritty stony scree, top-dressed around the vulnerable collar, winter wet will probably not harm it. It is, however, worth the trouble.

SYNONYMS: *C. allionii* Lapeyr.; *C. barbata* Lapeyr.

C. affinis Roem. & Schultes

This is a closely related and easily confused plant; it comes from a similar area, though more common to the south of the range of *C. speciosa. Flora Europaea* describes this as coming from the mountains of Eastern Spain, whilst its subspecies *bolosii* (Vayr.) Fedorov, also referred to in the past as *C. bolosii* Vayr, has its home on Montserrat. These two tend to have more open, flatter corollas.

There is a good modern description and illustration of *C. affinis* in the *Botanical Magazine* at t. 9568.

C. TAKESIMANA Nakai 70 cm × 40 cm Pale pink, Pale blue July/Sept

This is a plant which, described in a *Flora of Korea* as long ago as 1922, was more recently collected and introduced from the island of Ullung-Do, in the Sea of Japan off Korea. Careful publicity moved it on to the NCCPG 'Pink

List' of plants rare in cultivation; such is its habit that it has rapidly romped off this.

A strong, creeping rootstock throws up at intervals rosettes of flourishing, fleshy heart-shaped leaves, often with a reddish tinge, borne on a petiole usually much longer than the blade itself. This petiole is lightly winged, and also toothed at intervals. The whole is a bright glossy green, and the leaf-surfaces are markedly veined. Strong stems, held upright or slightly angled, bear a few more or less sessile, but similarly shaped leaves, and, in the leaf-axils and at the tips, long tubular bells of a pale greyish-pink with maroon inner markings reminiscent of *C. punctata*, to which it is no doubt closely related.

This is a handsome plant preferring a moist open soil, sandy or peaty; in this it is inclined to be rampant, and will consequently require some supervision. In our clay we find it well-behaved (or even inclined to die out, as does *C. punctata*) and, although it sets seed, it does not scatter this dangerously. In any case, so far as propagation is concerned, it is hardly needful to gather seed, as pieces of the running root, preferably with a node and rosette, are very easily established.

Further collections and introductions of this plant may well show that it is variable in form.

C. THYRSOIDES L. 30–70 cm × 30 cm Yellow July/Aug

This is one of the few campanulas which is never blue. Its other immediate distinction is its habit—a thyrse, being a closely-branched flowering stem with many short-stalked flowers. This thyrse is formed of a much longer cluster of flowers than seen in *C. glomerata*. The basal rosette is of wavy oblong to lance-shaped leaves. The stem rising from its centre is stout, erect and unbranched, with closely packed very bristly lance-shaped leaves; sometimes there is more than one stem from the rosette, but usually one is dominant. These stem-leaves become shorter higher on the stem, to the point of being but bracts in the axils of which arise the flowers; all so close as to form a compact steeple.

Individual flowers are with only slightly reflexed petal lobes. The calyx teeth are very narrow and there are no appendages. There are three stigma, and an ovary in three sections.

The colour is said to be yellowish-white, or pale straw, but there is a subspecies, ssp. *carniolica*, where the flowers are not so compacted together that is, the internodes are longer, the bracts are longer, and indeed, the whole inflorescence and stem are longer—in cultivation up to 1 m (3 ft). And in this subspecies the flowers are often of a better yellow. The claims of the species to our attention may be summed up in the concise assessment of an American lady visitor 'Well, it sure is interesting!'

35 *Campanula thyrsoides*

C. thyrsoides is seen at its best planted in close groups in the border, where it is utterly hardy, and where it sets seed abundantly. As it is biennial in habit, seed is the only way to propagate, and this is easy.

Mountain visitors will have seen *C. thyrsoides* in the high meadows from the Jura through the Alps and into the Balkan mountains.

The plant is portrayed in the *Botanical Magazine* at t. 1290, named *thyrsoidea*.

C. t. ssp. *carniolica* (Suend.) Podl.

Is found particularly, as the name suggests, in the Carniolican Alps of Austria and Yugoslavia.

C. TRACHELIUM L. 45–90 cm × 30 cm Blue-purple, White July/Aug

Bats-in-the-Belfry is a native, but not common, campanula in the wild in Britain. It is scattered in woods and hedgerows through the Southern UK and Europe. (It is replaced in Northern Britain by *C. latifolia.*) 'Nettled-leaved' is rather more descriptive, and the term that Farrer used.

The rootstock is woody, and the rough, acutely angled stems often emerge from a single base, a limiting factor when it comes to division of the plant for propagation. The leaves are bristly, and nettle-shaped, but happily they lack the sting. The lower leaves are on long petioles; those higher on the stem are shorter, and the leaves smaller. The flowers are tubular bell-shaped and slightly hairy, with the lobes dividing to about one-third. The calyx lobes are linear to lance-shaped, and there are no appendages. The seed capsule is round, nodding, and opens at the base. The flowers are rather like those of *C. latifolia* but not quite so long.

This may be thought too coarse a plant for the formal garden, but it looks well naturalised in grass or light woodland, and is not invasive at the root, though it does produce a lot of seedlings. Propagation by seed is easy, and can result in light blue or white. The white form is attractive as its flowers show up rather better against the leaves. An alkaline clay soil which is slightly moist seems to suit all forms best, but it is not essential.

C. t. 'Alba Flore Pleno' 60 cm × 30 cm White June

A delicious cottage garden plant, which shows off its semi-double white cup-shaped, fringed flowers to perfection against the light green leaves. It is quite tall and strong, and can usually be relied upon to flower very freely. Division is the only method of propagation, and it takes some courage to lift the plant and slice through the woody stem with a knife; this should only be done in spring as new growth commences.

C. t. 'Bernice' 40–60 cm × 30 cm Lilac-blue June

This double *C. trachelium* has been known since the late sixteenth century. The flowers are double amethyst-coloured. The delightfully fringed cup-shaped bell gives the impression of a flounced crinoline. It flowers profusely for a long time, with the flowers held well above the leaves. It is quite tough and hardy, and can be propagated by division, with care, as it is slow to increase. It will sometimes come true from seed. It was re-introduced by Alan Bloom, who received it with the name *C. lariafolia* which seems invalid. Its only faults are a slightly stiff habit, and an attraction for blackfly.

A bicoloured double form has been illustrated in old accounts, but we have never seen it.

SYNONYMS: *C. urticifolia/C. urticaefolia* Schm. (It should be noted that this

36 *Campanula trachelium* 'Alba Flore Pleno'

name has also been applied by different writers to *C. bononiensis, C. latifolia* and *C. rapunculoides,* all of which of course have nettle-like leaves!)

C. 'VAN HOUTTEI' 45 cm × 30 cm Violet-blue July

This is an old hybrid, which is often mentioned in books on *Campanula.* It is now only just in cultivation, and we are not quite certain that the few plants which are grown are actually the real thing.

C. 'Van Houttei' is said to be the other half of *C.* 'Burghaltii'—a reverse of the cross. That is to say it is a cross between *C. latifolia* and *C. punctata* with the former being the seed parent. It follows *C. latifolia* in the colour of the flowers, which are deeper than those of *C.* 'Burghaltii', being variously described as an indigo-blue, or a lavender-blue.

The leaves are long, oval to lance-shaped, with deeply notched margins. They are up to 10 cm long and 3–4 cm wide, hairy, with strong veins. The flowers are the same colour in bud and flower, are very large, up to 6 cm long, and hang in a drooping raceme. They have narrow pointed calyx lobes, which are about 2 cm long.

The origin of the plant is rather confused. It dates back at least to 1878, when it was said to have originated in France from Thibaut & Keteleer of Sceaux. However, other sources are said to be Dr Rodigas, and Messrs Van Houtte, of Ghent. Even the colour of the flowers is in some doubt, with some saying a deep indigo, and others a pale lavender. The plant we have reference to is a mid-campanula blue.

Robin Lane Fox described the plant thus: 'My pleasure comes from an inky mauve variety called Van Houttei. This is quite unjustly ignored. It is a bolder colour and flowers profusely. You can mass it in the front of a border where its heavy crop of tubular flowers draw the eye. I consider it a very fine border plant indeed which splits easily into 10s or 20s after a year.'[28]

Margery Fish spoke of the plant dying out with her, and the one grower who offers it says that it is not vigorous. It may be that there is more than one clone in cultivation.

C. VERSICOLOR Andrews 50 cm × 30 cm Violet-pale blue bicoloured July/Aug

Unlike the leopard, campanulas can change their spots. *C. versicolor* is reported enthusiastically by some, and much less so by others. E. K. Balls was an enthusiast: 'The stems . . . are thickly clothed with wide open stars, one and a half inches [4 cm] across. They are a fine purplish blue with a white ring and a deep reddish flush in the centre. Some of these aged plants on high cliffs produce as many as 20 of these beautifully curved wands of bloom.'[29] But in cultivation in ordinary border soil it grows upright, and can be rather like a dull *C. pyramidalis.* It was discovered by John Sibthorp in the late eighteenth

37 *Campanula versicolor* in *Flora Graecae* , Sibthorp and Smith

century, but was not in general cultivation until the 1930s. It received an AM in 1932.

The root is fleshy at first, but becomes woody and large as a fist, often wedged tightly between rocks in the wild. The basal leaves are smooth, leathery and olive-green, oval to oval-heart-shaped, notched, on long petioles. The upper leaves are on short petioles, and narrower. Several leafy stems end in spikes of flowers. In the best plants these are clearly bicoloured, but some seedlings have less distinct colouring. They are wide saucer-shaped, about 3 cm in diameter, and with spreading lobes. The style projects beyond these lobes, and the calyx lobes are narrow and reflexed. *C. versicolor* smells of cloves—one of the few campanulas with perfume.

The plant sets abundant seed which falls from the valves of the erect capsules near the base, and seed is the best means of propagation. The species is very variable in the wild, and plants from seed need to be watched for the best forms.

C. versicolor is a mountain plant from Western Greece, Albania, the Balkans and south-east Italy. E. K. Balls talks of it at 2,500 m (8,000 ft) in the Pindus range in Greece, where it 'flaunts masses of its lovely purple-blue flowers'.[29]

C. × *pyraversi*

This plant is more perennial than *C. pyramidalis*, which it resembles in form and habit; and it crosses with it to give what has been called *C.* × *pyraversi*, which is neither one thing nor the other, and does not appear to have anything in the way of virtues. It is suspected that plants offered as *C. versicolor* are, in fact, of this origin.

C. VIDALII H. C. Watson 46 cm × 30 cm Pink Aug/Sept

The campanulate flowers and seed capsules reveal *vidalii* as a *Campanula*, but it really is a very strange one, and the only shrubby member of the genus.

From a single woody main stem there branch many thick, scarred stems reminscent of a *Euphorbia*. The glossy, linear leaves are notched and have indented veins. They form shining regular rosettes rather like a succulent, and these elongate to become the flowering spike. The 3 cm bells are nodding on recurving stalks. They are a subtle dusky pink, and are waisted at about two-thirds their length. The waxy texture and waist give them rather the air of an Edwardian lady with an hour-glass figure. Inside the base of the flower the flat top of the ovary can be seen as a dramatic shining orange disk.

The stems do not branch again after flowering, and need to be cut out as they die off. When cut the plant gives off a milky sap, which dries to form a brown rubbery deposit. This can be disfiguring in the garden if the plant is wind damaged and, together with its tender reputation, it is probably better in a conservatory.

C. vidalii was first discovered in the Azores in 1842 when a Captain Vidal RN

14 *Campanula rotundifolia* in the wild

15 *Campanula speciosa*

16 *Campanula trachelium* 'Bernice'

17 *Campanula vidalii* at Kew Gardens, Richmond

found it by accident on a cliff-face near Santa Cruz. It was introduced by H. C. Watson. He described walking through Kew with a potted plant and bumping into Sir William Hooker. Challenged to guess what family it belonged to, Sir William guessed Proteaceae!

In 1872 Canon Ellacombe records growing it to 60 cm (2 ft) high at Bitton, and in 1903 a plant at Kingswear in Devon, which wintered outside, had 33 flower-spikes with 400 flowers in all.

It seems the plant can be grown outside in mild areas, and the National Trust at Powys Castle in Wales have used it effectively for summer bedding. Given a little heat and an open compost it makes an attractive plant in a conservatory or alpine house. As the flower-spikes form around the edge of the plant it is worth tying in to make a more compact shape. It suffers from red spider mite, and does not like root disturbance, but it is generally quite easy, and produces a lot of seed which germinates easily. (AM 26 July 1960.)

C. vidalii forma alba

The flowers are white, 4 cm long, with an orange ring at base of interior. The whole plant is slightly larger than the pink form, and they occur side by side in the wild. We have not seen this, and wonder whether it is still in cultivation.

SYNONYMS: *Azorina vidalii* H. C. Watson.

References

1. John Raven, *A Botanist's Garden* (1971), p. 210.
2. R. Farrer, *The English Rock Garden* (1918).
3. Ibid., p. 174.
4. Margery Fish, *Cottage Garden Flowers* (1961), p. 40.
5. Harold and Joan Bawden, *Woodland Plants and Sun-lovers* (1970), p. 64.
6. Geoffry Ely, *And Here is Mr. Fred Streeter* (1950), p. 147.
7. *Gardeners' Chronicle*, no. 51 (1897), p. 451.
8. Ibid., no. 56 (1899), p. 151.
9. Anon.
10. Robin Lane Fox, *Better Gardening*, p. 91.
11. David Stuart and James Sutherland, *Plants from the Past* (1988), p. 95.
12. Fish, *Cottage Garden Flowers*, p. 40.
13. Margery Fish, *Ground Cover Plants* (1964), p. 46.
14. Lane Fox, *Better Gardening*, p. 91.
15. *The Garden* (July 1901), p. 58.
16. Farrer, *English Rock Garden*, p. 187.
17. A. E. Gairdner, '*Campanula persicifolia* and its Tetraploid Form "Telham Beauty"', *Journal of Genetics*, vol. xvi, no. 3 (1926).
18. Alphonse de Candolle, *Monographie des Campanulées* (1830), p. 313.
19. Rev. Wolley Dod, *Gardeners' Chronicle* (Sept 1895), p. 335.
20. Gairdner, '*Campanula persicifolia* . . . "Telham Beauty"', p. 341.
21. Christopher Lloyd, *The Well Tempered Garden*, p. 84.
22. Tassel, *Gardeners' Chronicle* (1845), p. 224.
23. Farrer, *English Rock Garden*, p. 193.
24. *Gardeners' Chronicle* (Nov 1901), p. 328.
25. *The Garden* (1918), p. 129.
26. Farrer, *English Rock Garden*, p. 193.
27. William Sutherland, *Handbook of Hardy Herbs and Alpine Flowers*, p. 182.
28. Lane Fox, *Better Gardening*, p. 92.
29. E. K. Balls, *Gardeners' Chronicle* (Apr 1938), p. 283.

The Alpine and Rock-Garden Plants

The genus being essentially one from the mountains and high places of the northern hemisphere, it follows that the list of species which fall into the category 'alpine and rock-garden plants' in catalogues, and therefore in gardens, is long. The diversity of these is also great, from both the botanist's and the gardener's point of view. As we are unashamedly adopting the latter, the scope is somewhat diminished. Campanulas may be divided into three categories; these may be summarised as follows:

first, alpines for any and every garden
second, those for the specialist rock-gardener
finally, those which are a challenge to grow well, and to keep, for the most experienced plantsmen. (Some of our ablest alpine gardeners are in this field.)

Of these three categories, this account concerns itself principally with the first, though we hope that growers of the other two may also find help. This is not a comprehensive account of the genus: that must await another occasion. In numerous cases, species may be grouped; Crook, Farrer and others set the example here. Thus we deal with *C. rotundifolia*, the common harebell or bluebell, as a group, even though *Flora Europaea* may break it up into some 25 species, without counting those which could probably be added from the vast stretches of Asia. There are also a few from North America. It is certain, anyway, that those which have been in cultivation for any length of time will have cross-fertilised, and so lost any distinctive characters which may originally, in the eyes of a botanist somewhere, have justified specific status. The observant and enthusiastic gardener or plantsman will be content to note his plant as 'a good form' in colour or habit or length of flowering, which he will then propagate vegetatively in order to keep it and to share it.

We have therefore, dealt with the alpines more briefly than the border varieties. In choosing, we have selected species and cultivars obtainable in commerce. They may not, it is true, be found on the 'alpines' display of every Garden Centre. They can be looked for either as seeds or plants in the lists of specialist nurseries. Seed is also offered to Members by several of the specialist plant societies (see p. 138)—the *Campanula* list is often among the longest. The

38 *Campanula cochleariifolia*

plants may also be found at the sales tables of their meetings and shows. Remember though that this is a bit of a lucky dip. The name will only be as accurate as the records of the Member who collected and sent it in.

Cultivation

The special needs of each species are mentioned in the species descriptions, but a word or two should be devoted to cultivation and to propagation in a general sense.

A well-drained soil is the repeated *sine qua non* for all alpines. Whether it is a matter of water-drainage rather than porosity to oxygen, remains an unsettled question. However, it is probably both; a gritty, porous soil is certainly beneficial, which does not have to be impoverished by this. The ideal—largely unattainable to most of us, but still to be striven for—is a good loam with sand and grit in it. As an indicator, we use a John Innes-type compost* with up to

* John Innes base is seven parts sterilised loam, three parts peat, two parts sharp sand, with limestone and fertiliser according to the plant's stage of development.

an equal volume of 50:50 sharp sand and 6 mm ($\frac{1}{4}$ in) grit, for pot work and for propagation, and as a standard to aim at in the open ground. This said, it must also be emphasised that most campanulas are accommodating and easy of cultivation.

Many alpine bellflowers are of creeping habit. A few may even be troublesome in this respect. We recognise that such species owe their success in nature to this characteristic, witness *C. rapunculoides* among the more rampageous. It has been suggested from garden evidence that these sorts seem to exhaust the soil of some essential nutrient; they do not sicken it, nor are other genera in any way impeded subsequently. Unfortunately, investigation, chemical or otherwise, is not likely to be undertaken, as it would doubtless be very costly, and nobody would stand to make a fortune out of it! However, in practice, trespassing campanulas are not a great problem, and as a rule of thumb it may be said that the part which spreads is the part to take for propagation.

Alpine campanulas like an open sunny aspect. Once well established they can take considerable drought, and prolonged frost and cold will not harm them; their good humour under these two stresses depends upon a good root-system. It must be added that many are completely deciduous, dying back so completely in winter that many have inadvertently suffered at the trowel of enthusiastic gardeners of the tidying-up strain. If it is worth keeping, it is worth marking; more, with a label that the blackbirds will not scatter around the bed. If seed is not wanted, many small campanulas, like their larger sisters, may be given a haircut, when they will flower again, albeit less freely.

Propagation

The species to be discussed may all be increased by sowing seed in winter or early spring. Most will benefit from some freezing. They may be sown in small pans or trays or, if only required in small numbers, in shallow plastic pots. As mentioned previously, a John Innes sowing compost with the addition of an equal amount of a mixture of grit and sharp sand should be firmed, levelled, and the seed sown sparingly on the surface—it will probably be very fine. A further light dusting of grit over this should be made, and the container stood to soak in shallow water until the surface is seen to be dampened. The tray should then be placed in a cold frame, preferably north-facing, where it will be protected from drying out, from excessive rain, and from the depredation of birds. If the bed in the frame is covered with a layer of fine grit, and watered with a solution of potassium permanganate (tinted pink only) slugs, snails and worms will be discouraged. Containers should not be covered, as most campanulas germinate better in the light.

Seedlings may be pricked out when quite small, especially if they have been sown in a shallow container; with taprooted sorts this is preferable. If, however, the seed has been sown thinly, so that resulting seedlings are not overcrowded, they may be left to be transplanted with a good rootball when established. It is best to wait until the plants are filling a 7 cm ($2\frac{3}{4}$ in) pot before planting out in the open—they will then be ready to face the slug and the snail with less anguish.

It is easy to save one's own seed, and this will usually be abundant. The essence of the matter is the timing, as once the seed capsule has ripened in the open, and the pores opened, any slight breeze will soon scatter the contents beyond retrieval; campanula seed can be very, very fine. Removing the capsules with a piece of stem, and completing the ripening under cover, will save disappointment.

Increase may also be made by division in spring, or by taking side-shoots as they start to grow, when they will root easily. Any specially desirable forms as well, of course, as named cultivars, must be propagated by this method. As in so many other ways, campanulas will be found to be quite accommodating, and any gritty soil or compost will encourage root formation. We have found also that, at less favourable times of the year, an actively growing shoot with flowering buds removed will often root over bottom-heat. An ambitious mist system is not demanded; a daily fine spray where direct sunlight is not fierce will suffice to maintain a tissue–water balance until root formation is adequate. Well-rooted plantlets may then be potted and moved on as required.

Descriptions

In this section we provide brief descriptions of the more readily available alpines, in alphabetical order, and making the groupings already referred to. (By 'readily available' it must not of course be taken that these plants will be found commonly on the display and sales benches of the nearest, or any, Garden Centre.) Indeed, most, even if they are quite well known, will only be obtainable initially from specialist nurserymen, from the seed exchanges of the various Societies, or from friends. Omitted from our account are annuals, the rarer and more challenging species, and those which by common accord are not worth the growing. It will be retorted that the content of each of these three categories is very much a matter of opinion; we cannot but agree. We would refer the reader to our comment made earlier that this cannot be a comprehensive treatment of the whole genus. However, a list is given on p. 133 of some of these, including primarily those found in recent editions of the *Plant Finder*, together with brief comment.

C. 'ABUNDANCE'

This is a hybrid between *C. rotundifolia* and *C. arvatica*, closer in habit to the latter by past descriptions and, as the name suggests, more abundant, if smaller, flowers. It was judged worthy of an AM in 1915, but it is doubtful whether it is still to be found in cultivation. It is mentioned here in the hope of its possible rediscovery.

C. 'ALASKANA' 50 cm Blue June/Aug

This is a large version of *C. rotundifolia* in all its parts. Botanically it is merged with that species, as is the somewhat similar *C. r.* var. *groenlandica*, referred to especially in American literature. These come, of course, from Alaska and Greenland respectively, although they are probably dispersed over wider areas than just these.

C. ALPESTRIS All. 12 cm Blue June

This comes from lime-free screes in the Western Alps. Fine underground runners slowly form a mat by throwing up tiny rosettes of longish, narrow leaves, often slightly folded in on the central veins and with wavy margins. The almost invisible resting bud of winter opens up in the spring, and the rosette throws up a central stem which bears one, but occasionally several, flowers of a good bell shape and mid-blue, large for the size of the plant and length of stem. The flowers are often held in a characteristic horizontal fashion, are somewhat narrowed at the base, and the five calyx lobes are narrow with sharp points and half the corolla length.

This is a sought-after plant, long-lived and not difficult in cultivation, if given very good drainage but plentiful watering during the flowering period. Although it occurs naturally on lime-free mountains, it does not appear to be too fussy in the garden. It may be grown satisfactorily in a pot, but, like many others, in order to keep it in character should be potted on regularly.

C. a. 'Alba'

Known, though uncommon, and less robust.

C. a. 'Grandiflora'

In circulation, but if this is correctly named the flower appears slightly clumsy for the size of the plant.

C. a. 'Rosea'

An undistinguished pink, tending to muddy.

The type was awarded an AM in 1987; *C. a.* 'Alba' received the same award back in 1930, while a form shown as *C. a.* 'Frank Barker' received an AM, also in 1930, but this appears to be lost today.

SYNONYM: *C. allionii* Vill.

C. ANCHUSIFLORA Sibth. & Smith 30 cm × 25 cm Blue Summer

This is one of the grey- or silver-hairy campanulas from Eastern Greece, occurring on limestone rock. A substantial taproot forms a flat rosette of lyrate leaves. In this respect it is similar to *C. rupestris*, *C. topaliana* and others, and Crook puts them together in a group. The rosette, which is decorative enough to be grown for its own sake, gradually enlarges, and after one to three seasons' growth (generally two) hairy stems arise and branch shortly to bear many blue tubular bells of a pale shade. There is often a taller central main stem and diffuse semi-erect side stems. The flower is also hairy; the calyx of triangular lobes with very small appendages. The style bears five stigmas.

This plant is very much more hardy than its origins would suggest, and we have seen it safely through winters of repeated frosts down to some −12°C (10°F). It prefers some shelter, however, preferably that of an alpine house, to give of its very best. It is satisfactory in a pot, provided that this is not too shallow—the taproot needs adequate accommodation.

Although the plant, like many others of similar origin, is monocarpic, dying after flowering, seed is set abundantly; the species may be kept going in cultivation by this means.

The following are mentioned by Crook as composing this group. Few who have cultivated them will quarrel with him, especially when, as is so often the case, they are offered wrongly named.

C. andrewsii A. DC.
C. betonicifolia Sibth. & Smith
C. celsii A. DC.
C. ephesia (DC.) Boiss.
C. hagialia Boiss.
C. lyrata Lam.
C. rupestris Sibth. & Smith
C. stricta L.
C. tomentosa Lam.
C. topaliana Lam.

We have here confined ourselves to the names endorsed by either *Flora Europaea* or P. Davis in *Flora of Turkey*, Vol. 6.

C. ARVATICA Lag. 12 cm × mat Blue, Violet June/July

This is a sound perennial from the north of Spain which slowly forms a close mat of tufts of small, toothed rotund leaves on short petioles. Stems of 10–15 cm height bear a few similar leaves, and are topped, usually singly, with pale blue, or violet stars up to 25 mm across for a long period in June and July. In nature the plant is invariably found growing among rocks; in the garden it will be happy in a rock-crevice or a scree of well-drained limestone chippings, or even in tufa, though this will, of course, restrict its spread and probably limit its life. In the scree, which is probably the simplest way of growing it, it will move slowly and last many years in the open. *C. arvatica* will do quite well in a pot, but will need repotting, and breaking up at the same time for propagation, every other year or so. This is no menacing runner. An AM was obtained in 1952.

C. a. 'Alba'
This form is occasionally offered. It is attractive, but markedly less robust and long-lived in our experience. AM awarded in 1937.

C. AUCHERI A. DC. 15 cm Blue June/July
There are a number of campanulas from the Caucasus and Armenia which are so similar that even in nature they are difficult to distinguish. In cultivation they have doubtless hybridised to the point where the gardener, albeit an alpine devotee, will accept Crook's verdict that they are quite indistinguishable, and Ingwersen's 'the confusing group . . . all grading into each other to a bewildering extent.' Peter Davis, in the *Flora of Turkey* is of similar opinion; we are content to be of the company.

When offered *C. aucheri, C. bellidifolia, C. saxifraga* or *C. tridentata* we suggest quiet acceptance of the label, provided that the plant has dense rosettes of basal, spoon-shaped leaves, more or less rotund, with or without slight teeth at their tips ('tridentate'), the rosettes gradually building up to a rounded cushion which looks particularly dead whilst dormant. On close examination, these reveal the tiniest green bud which, amongst the first of the campanulas, revives to flourish and to throw up stems bearing open cups thrust to the sun, usually a deep blue with a pale or even a white 'eye'. A happy plant will conceal all foliage beneath the flowers, held singly at the tip of each sparsely leaved stem. The three-part style is shorter than the corolla. The calyx lobes are triangular-pointed, and the much smaller appendages also triangular. These are all lime-lovers or indifferent; alpine plants which may be grown equally well in pots, sinks or troughs, provided drainage is good, or in a hole in a tufa rock. Like most of their race, they appreciate a good collar of gravel or grit to protect them from winter damp, under which conditions they will prove to be trouble-free and long-lived.

C. aucheri, as such, was awarded an AM in 1960; *C. tridentata* obtained an AM in 1935.

C. a. 'Quarry Wood'
Obtained an AM in 1965. We hope that this form is still in cultivation, but have not been able to trace it.

C. 'AVALON' 20 cm × 30 cm Violet July
This looks like a *C. carpatica*, and is not infrequently quoted as such. It was, however, and is when found, a cross between *C. turbinata* and *C. raineri*. It is clearly intermediate between these two parents, of which the latter is the pollen parent (as opposed to *C.* 'Pseudoraineri', where the seed parent is *C. raineri*).

These two parents are closely related, and readily hybridise either way—all too readily, in many situations!

C. 'Avalon' has the habit of a neat *turbinata*, forming a close, tidy hummock of hairy stems and leaves; the flower is the open and full cup-shape of *C. raineri*. The leaves are mid-green, heart-shaped and pointed, with somewhat crinkled edges; the underside of the leaves and the long petioles are especially characterised by the fine stiff hairs. The flowers are borne as if resting on the cushion of foliage, which gives a particularly neat and beautiful effect.

This plant is slow to spread and not particularly rich in offsets, which is surprising with its two parents, but this could be the main reason why it is not common.

C. BARBATA L. 30 cm × 10 cm Blue, White Summer

This is a distinctive small campanula of the grass upper meadows of the Alps, also occurring in south-west Poland and one small area in Norway. It only

39 *Campanula barbata*

occurs in nature in lime-free soils, and is no doubt happier—which means longer-lived, in practice—in the same in cultivation; otherwise it tends to be somewhat short-lived, or has at least that reputation. Being also a meadow plant, it shows evidence of being happier grown in company, looking well in a raised bed with tufts of miniature grass like *Festuca ovina* 'Glauca' or *F. glacialis*. A perennial taproot or stolon produces a tuft of lance-shaped leaves in spring, and this throws up one, or several, stems when established, bearing but a few strap-shaped leaves and one or a few flowers, pendent and lavender-blue. Deeper blue is rare, but white is not infrequent, much more so in cultivation than in nature. These flowers have lobes divided to one-third, reflexed to reveal the hairs within which form the beard of the specific name. The style bears three stigmas; the calyx lobes are triangular, with short, rounded, reflexed appendages.

The taproot demands a deep soil, but this should for preference not be rich. A deep pot, as opposed to an alpine half-pot, can produce a good plant which will last for several years. Seed is usually abundantly set, and may be collected and used for propagation.

An AM was obtained in 1951.

C. 'BIRCH HYBRID' 15–20 cm Blue June/Sept

This plant, introduced by Ingwersen, was initially shown as *C.* × *portenscharskyana*; it was awarded an AM, with a recommendation that the name be changed. It is *C. portenschlagiana* × *C. poscharskyana*, fearsome names enough before hybridisation. The name 'Birch Hybrid' was resorted to after the nursery name, Birch Farm. In fact the use of the word hybrid is frowned upon under nomenclature rules, and *Campanula* 'Birch Farm' would probably be a better name. This is a strong-growing small campanula with a particularly long flowering period, and rightly very popular.

Strongly toothed kidney-shaped basal leaves on long petioles throw up branching stems bearing an abundance of pale mauve bells with spreading lobes divided to one-third their length. The style with its three stigmas is shorter than the flower tube.

Crook refers to this plant as similar to a major form of *C. portenschlagiana* in cultivation. In general it is a tidier plant, and, although claiming affinity with *C. poscharskyana*, certainly does not have the long branching of that species, nor the invasive propensities. It behaves well in a pot, but will need frequent moving on; breaking off the numerous rooted stems at the same time will give abundant progeny.

C. CALAMINTHIFOLIA Lam. 3 cm × 40 cm Lilac Summer

This plant is all too often confused with *C. sartorii*, the latter taking its place in the trade. It is at its best a short-lived perennial, whilst *C. sartorii*, which as a plant or grown from seed it often turns out to be, is quite relentlessly biennial (and sets seed and sows itself much more prolifically). A tiny rosette of slightly toothed heart-shaped leaves puts out ground-hugging, decumbent stems up to 20 cm or so long, profusely furnished with small rounded sessile leaves, and with upright slightly funnel-shaped lilac bells. The whole plant is greyish with fine hairs—a sure mark of its origins, which are the Greek Aegean islands. The style is longer than the corolla and ends in three stigmas. The calyx is made up of five triangular lobes, and the triangular appendages are small and insignificant.

This little plant will survive an average hard English winter outside, so long as it is planted in a very well-drained position, but is seen at its best in an alpine house. It is a lime-lover, and looks particularly well clambering over tufa rock; it is equally content in a limestone scree.

C. calaminthifolia is related, and similar in habit, to *C. heterophylla*, but this latter has distinctly spoon-shaped leaves, and is less hairy and grey.

SYNONYM: *C. sartorii* Boiss. & Heldr.

C. CARPATHA Halacsy 20 cm × 25 cm Blue Summer

This plant, little known or grown until recently, comes from the island of Karpathos, to the north-east of Crete, where it grows in shady, rocky places. Like most of the Greek mainland and island species, it is a lime-lover.

A tuft of long spoon-shaped darkish green and slightly felted leaves, strongly veined, round toothed on the margins, and on short winged petioles, throws up several stems which may be upright or more or less prostrate. These bear similar sessile stem-leaves and, in their axils as well as at the tips, tubular bells of a rich blue or blue-violet. These are cleft to about one-third, and the lobes of the 2 cm long flower only recurve slightly. There are five stigmas; the calyx lobes are triangular; the prominent appendages are roundish.

C. carpatha is biennial or monocarpic, and though it consequently dies after flowering, it usually sets seed well, and so may be easily propagated by this means.

Two species from Crete are similar to this campanula.

C. tubulosa Lam.

Occurs in the western mountains of Crete on limestone rocks and also in dampish places. It is hairier than *C. carpatha*, giving the leaves a darker and greyer appearance. The stems tend to branch more, the flowers are a little longer and they are of a paler lilac shade.

C. pelviformis Lam.

Occurs in the more easterly area of Crete; it appears as a smaller form of *C. tubulosa*.

Both *C. carpatha* and *C. tubulosa* have an AM to their credit, received in 1952 and 1933 respectively.

C. CARPATICA

C. carpatica has been dealt with under border campanulas (see p. 27); the dwarf forms, mostly derived from the subspecies *turbinata*, with a single flower to a stem, are both ideal and easy for a sink or trough. Careful treatment will produce good pot-plants also—this use of them is popular on the Continent—but they will need frequent repotting, and also some feeding, to be a success. Suitable cultivars include *C. c.* 'Hannah' and *C. c.* 'Karl Foerster'.

C. CASHMERIANA Royle 15 cm Blue July/Sept

C. cashmeriana is an exception to the general rule, which is that Himalayan campanulas are of little garden worth. To the enthusiastic grower of mountain plants who is familiar with the great range of wonderful plants from there, this may come as something of a surprise. We have observed, however, that environments which favour primulas are not those chosen by campanulas. *Campanula* and *Dianthus*, however, are soul mates. These generalisations also transfer to the garden; here, however, we can manipulate micro-climates within relatively small areas, to the advantage of many sorts of plants which would otherwise choose very different conditions.

C. cashmeriana is a small, wiry deciduous campanula which springs each year from a woody rootstock thrust by choice deep into narrow rock crevices. Fine

40 *Campanula cashmeriana*

stems covered with a white pubescence are held erect or trailing and sparsely furnished with elliptical, grey-hairy leaves. The stems are lightly branched above, and each terminates with a solitary pendent bell of aristocratic shape; they are of a grey-blue shade, again because of their fine hairiness. In a good clone they will be some 15 cm long and wide, which, as they are profusely borne on a relatively small plant, makes for a fair show. To this must be added that the flowering period is as long as that of any campanula in cultivation; each year we have them, in a sheltered spot, from July up to and into the frosts, which do not seem to worry them unduly. Seed is set with moderate enthusiasm, and they will self-sow, which is to their credit, as the seed is exceedingly fine, and not at all easy to gather by hand.

We have a soft spot for this little campanula, which is not nearly so often seen as it should be, for it is not hard to grow. However, we have to add that there are poor forms about; it is said that it hybridises with some of the poorer Himalayan campanulas (there is a multitude of names covering annuals and short-lived perennials, all of which resemble *C. cashmeriana* to a greater or lesser extent).

C. cashmeriana was awarded an AM in 1958.

C. pallida ssp. *tibetica*
Not dissimilar from the type, and could almost be taken for a dwarf version of *C. cashmeriana*, if one needed such.

C. CESPITOSA Scop. 12 cm Blue July

This species is very similar to *C. cochleariifolia*, and all too often, especially from seed, one receives this latter in its stead. The principal differences are as follows. *C. cespitosa* is a tufted plant, arising from a single taproot, and with no tendency to spread by underground runners as does, so generously, *C. cochlearifolia*. *C. cespitosa* tends generally to be somewhat taller in the stem, and the flower is usually pinched in at the mouth. Thus the overall impression is of a less dainty plant. The blue is variable, but mid-blue is by far the most common. *C. cespitosa* is much less widespread in nature, being found only in limestone screes and among rocks in the Eastern Alps and in the mountains of Yugoslavia.

Propagation must be from spring cuttings, or from seed, and cultivation is quite straightforward.

C. CHAMISSONIS Fed. to 20 cm Blue June/Aug

This is a sound, easily grown and long-lasting species which occurs over a wide area which includes Japan, Siberia, Sakhalin, the Kuriles, Kamchatka, the Aleutians and Alaska. This is a wider area of the globe than we are wont to realise; variations in the forms of this plant are also consequently great.

C.c. pilosa, dasyantha, chamissonis and other close relatives are described in the literature, and there is considerable overlapping and difference of opinion, but we note that both Kew and the *Flora of the USSR* give *C. chamissonis* priority; we accept therefore this name as covering what to the alpine gardener and enthusiast will be forms of the same species. As they all require the same cultural treatment, our task is easier.

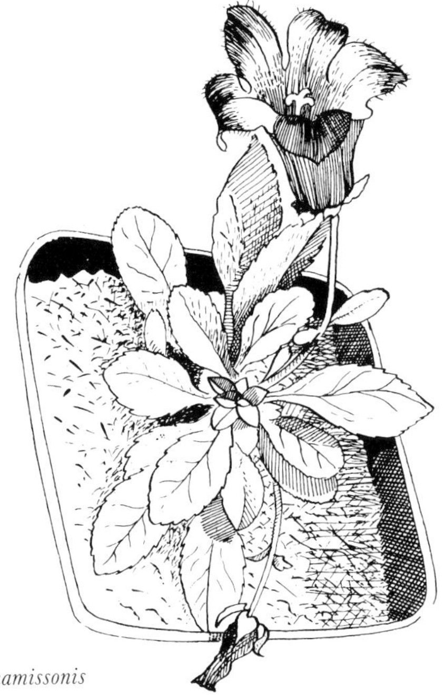

41 *Campanula chamissonis*

This plant has a stout perennial root which produces a rosette of usually evergreen smooth glossy spoon-shaped leaves with delicate and clearly marked veining. The margins are finely notched. These rosettes spread by underground runners, and the maturer ones among them throw up stems furnished with a few similar but more rounded, and stemless leaves, bearing at their tips long or rounded upright bells. These may often be very large for the size of the plant, and they vary in their shade of blue, frequently from the outside to the inside of the corolla. The variation in colour may give the impression that each petal is striped, sometimes quite dramatically so; frankly, we prefer less drama! This is a sturdy, easy and attractive plant. In nature it grows more commonly on lime-free soils, and in cultivation will flower better if we follow this taste, though it is not fussy. There will always be, in a mat of *C. chamissonis*, a number of rosettes which do not flower, though all will eventually. Grown in a peaty compost with good drainage provided by plentiful grit, this subject spreads slowly and happily, but never outrageously.

The Japanese have made this species a florist's flower, and have given the different forms a series of names, often in the manner of honorific titles. In Europe we find forms labelled as 'Major', 'Superba', etc., and these are frequently offered as such in garden centres. In themselves they mean little, and although such names have been correctly conferred initially, they have tended to rub off on all comers.

SYNONYMS: *C. dasyantha* Bieb.; *C. pilosa* Herd., Bieb.; *C. altaica* DC.

C. COCHLEARIIFOLIA Lam. 10 cm Blue, White June/Aug

The 'Fairies' Thimbles' which runs happily in crevices, screes, beds, sinks, troughs, paths, paving, drives, walls, and pretty well anywhere else, creeping and seeding joyously. Some consider it a menace; others despise it. It is both the easiest and daintiest of all campanulas; the world would be immeasurably the poorer without it.

The cause of the 'problem' is the creeping and branching rhizome which throws up tufts of small round or heart-shaped leaves, and stems with narrower leaves, topped by the 'thimbles' which vary considerably both in form and colour. In nature the commonest shade is a mid-blue, but white, grey-blue, China blue and lilac are known. Farrer found and named one 'Miranda', and generously predicted that it was going to be one of the greatest of rock-garden plants! It has pale bells on longer stems than the average, and one of its virtues is the length of its flowering period, from summer right through to autumn. 'Miss Willmott', selected in the garden of that lady in Warley, Essex, is a miniature mob-cap with slightly paler fringes.

C. c. 'Blue Tit' (China blue), *C. c.* 'Cambridge Blue' (pale blue), *C. c.* 'Silver Bells' (silver-blue)

These and a multitude of others are widely offered, and there are many flowering generously over a wide period. Looking again at nature, we could learn that this is a plant of rock-crevices and poor, gritty ground; it is both happy and restrained when we copy this in the garden. One of us grows it in a raised bed of nothing but gravel—and sand—in a mixture straight from the quarry; it, other campanulas, phloxes, including the Mexican ones, penstemons, together with a host of other alpine plants, grow well, flower well and persist, while showing perfect restraint in the spreading.

C. c. 'Elizabeth Oliver'

A double form of *C. cochleariifolia*; it is a very attractive powder blue. In general double flowers are not greatly admired by the establishment of the alpine world; here an exception should be made.

42 *Campanula cochleariifolia* 'Elizabeth Oliver'

C. c. 'R.B. Loder'
Indistinguishable from the above, receiving an AM in 1922 when shown by Prichards. We suspect that this is the correct name for the blue double, and we have yet to trace an authority for 'Elizabeth Oliver'.

SYNONYMS: *C. pusilla* Haenke; *C. bellardii* All.

C. 'E.K. TOOGOOD' 20 cm × 60 cm Violet Summer
This is a presumed hybrid, but of uncertain origin. It is like a large and lush version of *C. garganica*, and could well owe something to *C. poscharskyana*. It throws out robust leafy branches from a rosette of strongly toothed heart-shaped leaves, quite smooth and green, which bear bright blue star-shaped flowers along their length. The lobes of the corolla are deeply divided, and the base of the flower, the 'eye', is paler, almost white in colour. The style, which shares the white base and blue extremity, protrudes well out, and ends in three stigmas. The long-pointed calyx lobes reflex on to the ovary and the short stalk.

This campanula, which is quite freely available, makes a fine display especially growing on a wall; it flowers over a long period.

C. 'Constellation'

Very similar to the above. It has larger leaves which tend to be more pointed, and more heavily toothed; the flowers are also larger, but lack the pale base to the corolla. The calyx lobes are not reflexed, and the style appears not to divide at its tip; we suspect it is a mule, for seedlings are not found and we have not been able to collect seed. For propagation, of course, this is not important, more especially as division in spring is very easy, as are cuttings.

Neither of these two plants form runners, and so remain always manageable.

C. ELATINES L. 15 cm × 30 cm Blue Summer

This belongs to a group which includes the well-known *C. garganica*, and also *C. elatinoides*.

All have small round to heart-shaped leaves, usually not more than 1 cm wide. These are delicately toothed, and the basal ones on longish petioles, the stem-leaves having ever shorter stalks up the stem, which are semi-prostrate or rising. The flowers are finely-shaped stars the length of the stem on short pedicels. In general these are all neat, compact plants which do well either on the open rockery or in pots, being lime-lovers and weather-resistant.

In an endeavour to simplify the group, and at the risk of overdoing simplicity, we describe the following from among the plants which are commonly offered. The *Plant Finder* has been our guide in this as in many matters.

C. garganica

Has evergreen tufts of tiny, bright green finely toothed leaves; bright blue stars are held to the sun on short, fine stems and pedicels.

C. g. 'Blue Diamond'

Resembles the type in habit and size, but has a paler base to the corolla, which in an open flower appears as a sort of pale blue five-sided diamond.

C. garganica ssp. *cephallenica*

Comes from the island of Kephalonia off the western Greek coast. It is larger in all its parts; a robust plant, but probably not quite as hardy as the above.

C. g. 'Dickson's Gold'

Of similar habit and size to the type, but the leaves are distinctly yellow; gold in summer sun. This is a very appealing plant, but generally slow to establish, and not in any case as robust or long-lasting as the above.

C. g. 'Hirsuta'

A very grey-hairy form, with the same long and abundant flowering stems; the flowers are dusky blue due to the hairiness, which covers the whole plant. This sets seed abundantly, but in our experience the progeny are not hairy; propagation of the hairy form must be from cuttings in spring.

C. g. 'W. H. Payne'
Has a striking white base to the corolla.

C. elatines
More compact; the leaves are thicker and rounder, the flowers are not so showy. The whole plant is finely hairy, but not enough to be grey.

C. elatinoides
Very densely hairy and grey, with again the thicker leaves; it is now classified as a subspecies of *C. elatines*.

C. fenestrellata
Very evidently of this same group. The leaves are a bright green, larger than any of the foregoing, and heavily toothed. The stems of this plant tend to be slightly ascending, as opposed to the ground-hugging, procumbent stems of *C. garganica*. *Flora Europaea* helps here by saying that the pollen of *C. fenestrellata* is blue, whilst that of *C. garganica* is yellow. The observant grower will generally also note that *C. fenestrellata* has longer stems, with smaller flowers than *C. garganica*.

Most possible permutations of the names of members of this group have been made at different times by different authorities in the past.

The following names of this group are among the more commonly met, albeit with incorrect status. These are as found in *Flora Europaea*, being the most recent authority covering the geographical area involved:

C. elatines L.
C. elatinoides Moretti
C. garganica Ten. ssp. *acarnanica*
ssp. *cephallenica*
ssp. *garganica*
C. fenestrellata Feer ssp. *debarensis*
ssp. *fenestrellata*
ssp. *istriaca*

C. 'ENCHANTRESS' 20 cm Lilac Summer
We are not sure that this plant is still to be found in cultivation; as in the case of other award-winning campanulas, it is mentioned here in the hope of its rediscovery.

Obtaining an AM in 1918 when exhibited by Mr Grove of Sutton Coldfield, it is, or was, a cross between *C.* 'Norman Grove' (see p. 112) and *C.*

waldsteiniana. In habit and appearance it inclines towards the latter, being of tufted habit, some 15–20 cm high, the stems bearing small, narrow leaves and semi-pendent pale lilac-mauve starry flowers.

C. EXCISA Schleicher 15 cm Blue July-Aug
This species, although probably not amongst the easiest, nor very long-lived in cultivation, is none the less attractive and popular because, like *C. zoysii,* it is just that little bit different!

43 *Campanula excisa*

It is a rambler in the style of *C. rotundifolia,* and throws up among the rocks and screes in which it occurs thin wiry stems laxly clothed with narrow smooth-edged leaves terminating in single, usually pendent, tubular bells with outcurving pointed mouths, between the pointed lobes of which are the curious round punched-out holes which have given the plant its name. The stems are only about 8–10 cm in height in nature, but in cultivation tend to be somewhat taller.

This is a real rambler, and so not happy in a pot for long. The ideal spot would be a granite scree containing a little peat below, for, as Farrer remarked, although the best plants are found in sunny spots, there is always some moisture beneath. Semi-shade, or a position where the mid-day sun in summer is shaded, is a fair compromise, but it must also have room to travel; otherwise it will be very short-lived in cultivation. In our experience C. excisa should not be allowed to dry out completely in winter, but equally should not be allowed to receive an excess of damp.

C. excisa is only found in a quite restricted area of the South Central Alps, in the Monte Rosa, Matterhorn, Simplon Pass triangle.

Propagation is from seed and rooted runner-cuttings in spring. This species is always found on lime-free scree, and lime should be avoided in its culture.

C. FORMANEKIANA Degen & Doerfler

Although this has been discussed under Border Campanulas (see p. 32), largely for its potential height, it is likely to be of even more interest to the alpine gardener. It makes an excellent subject for growing in a pot where, if given alpine-house shelter, it makes a most attractive display over a long period.

C. 'G.F. WILSON' 10 cm Violet-blue July/Aug

This is *C. pulla* × *C. carpatica* ssp. *turbinata*. Slowly spreading tufts of slightly pointed oval leaves arise from underground runners. These leaves are of the slightly yellowish colouring which often betrays hybridity; as originally described in the literature, there were two forms of this cross, one considerably more yellow than the other and not apparently common, if in cultivation, today. It was said to be by far the less robust of the two. From each tuft of leaves arise sparsely leafy stems bearing similar leaves and topped by semi-pendent bold bells of a deep violet-blue. This is a most attractive plant, and further increases in value because of its later flowering, which is at its height in July and to the end of August, even into September.

C. 'HALLII' 10 cm White July

This is a cross between *C. cochleariifolia* and *C. portenschlagiana*, but betrays not a lot of the influence of the latter, though upon close examination it can be seen that the glossy pale green leaves are intermediate between those of the two parents. The white flowers are borne singly on the 10 cm stems; they are semi-erect and are more open than those of *C. cochleariifolia*, for the white form of which it could easily be taken. Had it not been raised in his Yorkshire garden by a Mr Hall, who was a noted grower of alpines, its identity would the more be called into question. It runs, but not as badly as *C. portenschlagiana*, the pollen parent.

This campanula obtained an AM in 1923.

C. 'HAYLODGENSIS' 15 cm × 10 cm Pale blue Summer
This is a cross between *C. carpatica* and *C. cochleariifolia*, raised by a Mr Anderson-Henry at Hay Lodge, near Edinburgh in 1885.

44 *Campanula 'Haylodgensis'*

Originally, this campanula was a single-flowered plant, with the habit largely of the seed parent, *C. carpatica*, but showing the influence of the pollen parent in the erect openness of the cup-shaped bells. We could not positively assert that this single form is still in cultivation as originally known, though it would, of course, be easy enough to repeat—and no doubt has been. However, the plant now well known under this name is a double of a light blue. The plant runs, but only slowly, and in fact its robustness and hardiness have been queried. It succeeds in a well-drained position, like most of its kin.

The flower of this campanula is larger and more open than that of *C. cochleariifolia* 'Elizabeth Oliver', which is also of much more recent origin, and they are borne on taller stems. The leaves are also larger.

A white seedling of this plant is referred to under *C.* 'Warley White'.

C. 'JOHN INNES' 15 cm × 70 cm Violet July

This is an unlikely cross between *Campanula carpatica* as the seed parent and *C. versicolor* as the pollen parent. As soon as one sees it the parentage is indeed evident, for it appears what it is—*C. carpatica* in habit with *versicolor* flowers. A tuft of long pointed heart-shaped leaves, intermediate between those of the parent plants, throws out radiating runners 30 cm or more long, slightly

45 *Campanula* 'John Innes'

branched and, typically, turning up at the ends. These bear, from July onwards, a succession of upturned star-shaped bells some $3\frac{1}{2}$ cm across, of a rich lilac or lavender with violet centres. These flowers, with their deeply cut and sharply pointed centres, have a lightness not seen in the *carpatica* tribe generally.

This plant is long-lasting and does well in any normal soil, and looks particularly good at the base of the larger rock-garden. Because of its long branching runners, it is not so suitable for pot culture.

This plant is sometimes found under the name *C.* × *innesii*; under nomenclature rules, 'John Innes' seems more appropriate. Crook in his monograph referred to *C. innesii*, and gave the parents as *carpatica* and *pyramidalis*, but we think this a slip on his part, as his notes and cuttings from past literature refer to *carpatica* and *versicolor*.

C. 'LYNCHMERE' 30 cm × 25 cm Violet July/Aug

This miniature campanula won an Award of Merit in 1948, and Crook reported it as being showy and full of promise. In spite of this it is not now in wide circulation. It is, however, obtainable, and invariably attracts the

46 *Campanula* 'Lynchmere'

attention of every observer when it is in flower. It is thought to be a cross between *C. elatines* and *C. rotundifolia*, though *C. cochleariifolia* has also been given as a putative pollen parent.

It forms a small 'bushlet' some 20–30 cm in height of fine, branching stems with bright green oval leaves with blunt but regular teeth and a rounded rather than sharp tip. The branching is sparse, and the terminal flowers are pendent bells of a rich violet-blue about 2 cm in length and 1 cm across, the slightly recurved petals of which are only shallowly divided. The colour is outstanding as revealing little of the red tone which is quite general in campanula flowers. This little plant is happy on the rock garden or in a pot, is long-lived, and, if given a haircut after flowering, will give a second crop of flowers until the frosts of autumn. Seed is not apparently set, but cuttings in autumn are as successful as those taken in spring; the former, perhaps with some bottom heat.

C. 'MIST MAIDEN' 20 cm White June/July

Although the origins of this are obscure, it is certainly well named. Brian Mathew, reporting its Award of Merit in the *Quarterly Bulletin of the Alpine Garden Society*, wrote of its being among the most graceful of the white-flowered campanulas. It was shown for its award, in 1981, by Ingwersens, but its raiser is unknown, as is the exact parentage, though it is not hard to see the probability of *C. tommasiniana* having a hand somewhere. Pure white pendulous or horizontal bells with flared mouths are held five or so to a fine wiry stem, making a well-grown plant seem very floriferous. Leaves at the base of the stems are oval to widely lance-shaped, about 1 cm long on equally long, or longer petioles; the few up the stems are narrow and the petioles become shorter.

This plant is suitable for rock-garden or pot cultivation, and in spite of its delicacy is thoroughly perennial. It spreads slowly by underground runners, and these may be used for propagation, preferably in spring as growth restarts.

C. MOLLIS L. 5 cm × 35 cm Lavender June/Sept

This rock-garden campanula comes from the south of Spain, but has close relations in neighbouring areas of North Africa. It has a somewhat confusing nomenclatural background, having borne the names of *C. malacitana*, *C. velutina*, which, added to the subspecies names accorded by the best of authorities, can leave one bewildered. *Flora Europaea* favours the name of *C. mollis* at least for the European species, and this is, as far as we can say, by far the most common in cultivation; we hold to it.

This plant, from a thoroughly perennial rootstock, forms a rosette of small spoon-shaped leaves. From among these are thrown out a number of ground-hugging stems of about 20 cm long, towards the end of which appear the flowers, upturned funnel-shaped bells of pale lavender or lilac (the shade can vary considerably) on longish pedicels. The stem-leaves are round and stalkless. The corolla is split to half-way and the petals are flared to show a paler centre to the flower; the veins, however, are stained lilac of a stronger hue. The whole plant tends to be grey with short silky hairs. It obtained an AM in 1932.

C. m. var. *gibraltarica*
The whole plant is larger in all its parts; it appears to be more floriferous; and we can affirm that in our garden, in open ground in a 'Mediterranean house', it has been accustomed to show at least a couple of flowers in every month of the year. The flowers are 2 cm long as grown by us; the corolla is pale without, but the inside is a deep rich lavender with pale base. The style bears three stigmas which are much shorter than the corolla, reaching barely to the base of the clefts of the lobes, which themselves are not so reflexed as in the above type. Calyx lobes are long-pointed, with small triangular reflexed appendages. The deep violet staining of the petal lobe veins remains, but is not so outstanding against the deeper overall shade.

This is a rewarding plant to grow, but, the prostrate stems being very fragile, wind and rain are their constant enemy unless they are sheltered. Like most Mediterranean plants in the British climate, sharp drainage is essential. The single rosette forbids division, but seed is usually set in the sheltered position recommended, and this offers the best propagation method; late winter sowing seems to give best results.

SYNONYMS: *C. malacitana* Herv.; *C. velutina* Desf.

C. 'MOLLY PINSENT' 22 cm Violet-blue July/Aug

This is a little-known and badly documented hybrid of not surprisingly unknown origin. A not-strongly spreading rootstock produces lax rosettes of crinkled heart-shaped finely pointed leaves which have a tendency to fold up from the midrib. They are longer than broad, and held on petioles about equal to their length. These leaves are irregularly and widely toothed, almost lobed in part. The sparingly branched stems, some 20–25 cm in height, bear similar but sessile leaves, and, singly at their tips, violet-blue flowers of a good rounded bell-shape. The corolla lobes are divided to about one-third, and but slightly reflexed; when opened they are held horizontally. The calyx lobes are long and fine, without appendages.

The leaf suggests the almost inevitable *C. carpatica* parentage; more is unknown, and the only, brief, mention we have traced is in Alan Bloom's *Alpines for Your Garden*, 1980.

We have this little plant in cultivation in Cambridgeshire; it flourishes in a rock-garden or sink, is hardy, easy and trouble-free.

47 *Campanula* 'Molly Pinsent'

It may be noted that Cecil Pinsent was a garden-designer in the 1920s, and there is a hybrid saxifrage 'Kathleen Pinsent', but we have no further detail of the original naming.

C. 'NORMAN GROVE' 15 cm Blue Summer

The origin of this hybrid is now uncertain, but *C. isophylla* and *C.* 'Stansfieldii' are given by Crook as the possible parents. If so, the habit more resembles that of the latter than the former. Certain is it that it is a robust and long-lived perennial, and reasonably hardy in a well-drained soil. From a tuft it spreads slowly into a mat of small triangular heart-shaped leaves from which ascend stems with few lance-shaped leaves and topped by semi-pendent cup-shaped bells of mid blue, borne on longish pedicels.

Norman Grove, of Sutton Coldfield, was a campanula enthusiast who grew and showed a number of alpines before the outbreak of the First World War. *C.* 'Chastity' was shown by him and obtained an AM in 1916. This was a free-flowering seedling of *C.* 'Norman Grove'; a plant a little taller, to 25 cm, bearing, as the name betrays, flowers of a pure white. Another of his awards plants is noted under *C.* 'Enchantress'. We have not been able to trace modern sources of either of these plants.

48 *Campanula* 'Norman Grove'

C. PORTENSCHLAGIANA Schultes 25 cm Blue June/Aug

This is the former *C. muralis*, a name which gives a hint to its best use, which is for mural decoration, that is, growing in or on a wall. It is one of the menaces, albeit a modest menace, among campanulas, in that, not content with the long questing stems reaching out to cover the ground, it has more reaching out below the surface of the soil in quest of fresh territory.

A fairly fleshy rootstock puts up a rosette or several rosettes of tiny heart-shaped leaves on very long stalks; these increase in size to about 5 cm across; they are heavily toothed, often multiply so, and the margins are usually wavy at full size. There is no point at the tip of the leaf. Stems up to 40 cm long are thrown out from among the leaves. These bear similar but smaller stalkless leaves, and in their axils branch stems which carry a number of flowers on short side stems. The flowers are funnel-shaped bells cleft into lobes about one-half their length, and these lobes are flexed outward about 45°. The style is a little shorter than the petals, and bears three stigmas. The calyx lobes are small and fine-pointed, without appendages.

The colour of the flower is a uniform deep lavender. A selected form, from the Continent and known as 'Resholdt's Variety' is larger in all its parts, particularly the flowers, but these are paler in colour.

This campanula does not need a rich soil, although it will do well in one. It will grow in crevices in the top or side of a wall with happy effect. Although the runners do quest far, in a more open situation they are not at all hard to control by just pulling them out; only in a neglected situation will they spread too far! These same runners, of course, provide all the propagation material required.

C. portenschlagiana received an Award of Garden Merit in 1927.

SYNONYM: *C. muralis* Portenschl.

C. POSCHARSKYANA Degen. 25 cm Various June/Sept

Although this resembles *C. portenschlagiana* in many respects, the most immediately obvious difference is the shape of the flowers, which are here star-shaped. It is not quite such a coarse plant; the leaves are shorter stalked

49 *Campanula poscharskyana* 'E. H. Frost'

and smaller. On close examination the following differences also appear: leaves with slightly more pointed tips; toothing finer; stems covered, albeit sparsely, with fine bristles; the leaves less so.

The flowers, as we have said, are quite distinct. Here they are cleft to three-quarters or even four-fifths the petal length, and the resulting lobes are widely reflexed to give a flat star-shape. In our experience this plant, in the forms mentioned below, is less of an invasive creeper than *C. portenschlagiana*.

Side rosettes or short runners, with or without roots, may be taken for propagation as the plants resume growth in spring. These will all flower quite early in the campanula season, and if shorn after blooming will crop again right into and even through the frosts.

An AM was awarded in 1933.

C. p. 'E. H. Frost'

A milky-white which shows itself off especially well in a sink or trough, or on the edge of a raised bed, where it will tumble over the edge.

C. p. 'Lisduggan'

Attractively pink; hardy as the type but not so strongly growing.

C. p. 'Stella'

A particularly bright blue, and the flowering stems can reach 40 cm or more in a rich soil, which, incidentally, is not necessary. It would appear that two hybrids mentioned in their place elsewhere are closely related, but with stronger growth still—*C.* 'E. K. Toogood' and *C.* 'Constellation'. These could have the admixture of *C. isophylla* 'blood'.

C. 'PSEUDORAINERI' 15 cm × 15 cm Blue July/Aug

Farrer brought this one up, and although he was, for him, curiously indecisive about the name, he seemed to think well enough of the plant. Anyone who has grown much *Campanula raineri*, and had our old friend *C. carpatica* within wooing distance, has experienced the disappointment of sowing the carefully gathered *raineri* seed, to find that it had been unfaithful. No matter; a fine plant results, and probably one that is easier to keep than the true, pure *raineri*, less attractive to slugs, able to freeze for days in a small pot without failing to make a comeback—in brief, a valuable plant. Of course, the lingering uncertainty of its murky ancestry will cast gloom and loss of esteem over the purist, but the fine plant for its own sake remains. Because it is no new-fangled invention, we here propose to fix the name of *C.* 'Pseudoraineri' for the issue of a union between *C. raineri* as the seed parent and *C. carpatica* ssp. *turbinata* as the provider of pollen. There will be some variation in this first filial generation, but really remarkably little.

A slowly creeping tufted plant—like both parents—produces small, round and slightly pointed heart-shaped leaves, between parents again, and sparsely leafy stems about 10–15 cm in height, with open blue saucers some $3\frac{1}{2}$–$4\frac{1}{2}$ cm across, rich and long-lasting in bloom. The chief variation may well be in the hairiness of the plants—hence the overall greenness or greyness, and in the shape and carriage of the petals.

The best can, of course, be proliferated by division, or cuttings in spring if a quantity is desired; those of the Irish variety can usually be found in plenty.

C. PULLA L. 12½ cm Deep blue June

This is one of the darkest coloured campanulas; *pulla* means 'very dark'. It comes from the Alps of Styria in Austria—a rich area for alpine campanulas. It is one of the best known and is freely available. It forms a mat of rosettes, each of small round shiny leaves on short petioles, and from virtually every rosette a fine stem about 10 cm high, with a few stemless round and pointed leaves, bears a single pendent well-rounded bell of a deep purple-blue. The mat extends slowly, and this plant is not by any means invasive; some would assert not enough so! It is ideal in a sink or trough, but seems to exhaust a pot rather rapidly and is not happy there. It is a lime-lover, not rapid to establish, and tends to flower itself to death if not frequently divided. It has been said to demand shade; others are equally sure it requires a sunny spot; we cannot detect much difference in results, but are certain that it does not tolerate drying out. A moisture-retaining soil, with abundance of both grit and humus, and due attention to water requirements, are the secrets, it seems, of success with this beautiful little campanula.

Propagation is by lifting a rosette in spring, preferably with some root attached, and replanting in a shady, moist compost until thoroughly re-established. It is one of the many campanulas which is completely deciduous, and disappears in winter. Many have thought it dead and gone, and, digging its position during the dormant season, have made sure it is dead and gone!

C. pulla received an AM when shown in 1976.

C. × PULLOIDES 15 cm Dark blue July

This is a garden hybrid—the two parents grow in nature some 300 km apart. As so often *C. carpatica* spp. *turbinata* is one, and the last entry, *C. pulla*, the other. We would expect the latter to be the seed parent, as our subject resembles it the more closely of the two. There is the root-running habit and the deep flower colour of *C. pulla*. The basal rosettes are sparsely furnished, and the leaves larger than in the latter; the stems bear few leaves, which are longer, and the stem height is greater—some 15–18 cm. The flowers are of the same deep violet-blue colour, but more pendent, larger, and somewhat 'blowsy' in their shape, as if composed of slightly crinkled paper. The flowering is a couple of weeks later than *C. pulla*.

This is not a robust plant; it needs an open gritty soil, and it should be renewed constantly. If happy, a mat of it, preferably on a limestone scree, is an attractive feature of a raised bed.

C. RADDEANA Trautvetter

C. raddeana, referred to in the Border Section (see p. 68), is also eminently suitable for cultivation in rock-garden, scree, sink, or even pot. Indeed—

though not by any means exclusively so—it makes a good beginner's rock-garden plant. It will grow truer to type, standing firm and strong, when grown in gritty and poorish soil, so long as this is not allowed to dry out excessively during any prolonged period of drought.

C. RAINERI Perpenti 8 cm Blue July

The most difficult aspect of this little treasure is finding the real thing! It comes from a very restricted area of the limestone Southern Alps in the Bergamo region, where it runs slowly in the rock and scree to throw up tufts of mid-green leaves, slightly but very regularly toothed in a quite characteristic manner, reminding one perhaps—though in this character only—of a *Dryas* leaf. Stems some 8 cm in height are thrown up, and these are terminated by an unexpectedly large open saucer of a blue flower.

In cultivation, the ubiquitous *C. carpatica* crosses all too easily with this little plant, so that, unless open pollination can in some way be prevented, guarantees go with the wind as easily as the bees carry the pollen. The result of this hybridisation is what has been called by us, and others before us, *C. × pseudoraineri*.

50 *Campanula raineri*

C. raineri is completely deciduous and quite hardy; it rejoices in a limestone scree which must be well-drained below. It creeps but slowly and is best in a sunny but slightly sheltered place where it will spring up each year without fail. Watering should be generous during growth and flowering, and the area should not be allowed to become too dry at any time. On the other hand, winter wet allowed to hang around will surely kill; as with all true alpines, drainage and root-oxygenation are all important.

C. raineri is not all that happy in a pot, or not for long without being moved on; this may be done in spring when growth restarts, and rooted underground runners may be taken off for propagation.

First find your plant; although *C. raineri* is perhaps a connoisseur's plant, it is not too difficult to cultivate, but unless the leaf is quite emphatically longer than its width, and only slightly and regularly toothed, one must be suspicious of an intrusion of *C. carpatica*. Seed from a dependable source is probably the best guarantee, unless a reliable pedigree is offered with the plant!

C. RHOMBOIDALIS L.

The same comments regarding use in rock-garden, sinks and troughs may be made for *C. rhomboidalis*, which *Flora Europaea* gives as a close relative of *C. rotundifolia*. As is so strangely, yet so often, the case, they remain more distinct to the gardener than to the botanist.

C. ROTUNDIFOLIA L.

This plant has been discussed under the border campanula section (see p. 73). It is eminently suited to the rock-garden, and the smaller members of what we have treated as the *rotundifolia* group may be grown very effectively in raised beds, and even more so in sinks and troughs; in any case, it does best in a poor gritty soil. For such use, we have had good experience of *C. carnica*, *C. justiniana* from Yugoslavia, and *C. recta* from the Pyrenees.

Among the most useful, and justly popular, of the group for the rock-garden is the beautiful deep-coloured *C.* 'Covadonga' first introduced by Clarence Elliott and Dr Roger Bevan from the north of Spain many years ago; this is also referred to in the border section under *C. rotundifolia*.

C. × STANSFIELDII Auct. 12 cm Violet-blue July

The parentage of this old hybrid is obscure, but it is thought to be a cross between *C. tommasiniana* and, perhaps, *C. carpatica*. It forms mats of pale green, hairy, somewhat diamond-shaped leaves and throws up stems of 10 or 12 cm in height and scattered on the spreading mat. These stems are topped each by shallow but wide violet bells, pendent and with corolla lobes divided to about one-third their depth and well reflexed. As well as being 'a gift of heaven . . . a treasure of the highest claims'[1] (Farrer extravaganza) this little plant has been the parent of other useful hybrids, including *C.* 'Norman Grove'.

Farrer suggests that this may in fact be a natural hybrid, and he puts forward the names of *C.c. tommasiniana* or *waldsteiniana* with *C. pulla* as the possible parents, occurring as they do at one point on common ground on Monte Maggiore; evidence in favour of this being his claim that batches of plants of these species (sent from there in days when such collections from nature were not frowned upon) contained some of this apparent cross. This must remain something of a mystery, albeit an interesting one, as, although the literature contains many mentions of our *C. × stansfieldii*, little is said which helps to solve the problem. Because we are inclined to accept the theory, however, we retain the form of its name given here.

C. TOMMASINIANA Koch 10 cm × 15 cm Lavender Aug

From the north of Yugoslavia hails this attractive little species. A thick perennial root throws up numerous branching wiry stems bearing lance-shaped and lightly toothed leaves and a mass of hanging somewhat tubular bells of pale blue. With us this is one of the later flowers, but it is not difficult in cultivation. It seems to thrive in any well-drained position, in sun or shade; it will be happy in a pot, a sink or open scree. It is a lime-lover. Runners are given off from the main root, and these may be used for propagation when taken off in spring, generally with some fine root-hairs already formed.

A characteristic feature of this plant is the almost horny tip to the leaf.

51 *Campanula tommasiniana*

C. waldsteiniana Schultes

A similar species in habit and appearance, but here the flowers are quite distinct; they are up-turned star-shaped bells, generally of a deeper blue than *C. tommasiniana*.

Both of these are long-lived plants with few foibles in our experience.

C. × TYMONSII 12 cm China blue Aug

This is a neat little hybrid between improbable parents, and quite different from the product of another get-together, *C.* 'John Innes'. The offenders are *C. carpatica,* of well-proven promiscuity, and *C. pyramidalis.* Whatever the skeletons in the cupboard, the result is a success on the rock-garden, for the evidence of *C. pyramidalis* is scant to the view. A slowly, but not menacingly, spreading mat of toothed oval and bluntly pointed leaves of glossy green on short stems which prolong to some 10–12 cm are thickly topped by open, shallow and chubby bells of China blue. This campanula is, in our experience, the latest of all to flower, and thus is all the more welcome in the garden.

This plant is beloved of the slug and the snail, and is best grown where these can be more easily controlled, such as in sinks and troughs; here a limestone scree provides ideal conditions for growth, and from this the runners may be gently pulled in spring for propagation.

52 *Campanula × tymonsii*

It would appear from older accounts that two other campanulas of the same or similar parentage were known—these are *C.* 'Fergusonii' and *C.* 'Hendersonii', but these seem now to be lost. Inadequate descriptions make the location of such plants almost impossible.

C. 'WARLEY WHITE' 15 cm × 25 cm White Summer

There has been some confusion over the naming of this plant, which has been called *C. warleyensis, C.* 'Warley Alba' and *C. warleyensis alba*. There is little doubt that all refer to one and the same plant. This campanula originated in the garden of Miss Ellen Willmott in Warley, Essex, a seedling from *C.* 'Haylodgensis'. There is on record a somewhat indignant letter from Miss Willmott, who had shown the plant at the RHS Flower Show and received no award, whilst a year or two later the same plant, shown by a nurseryman who also had a large trade exhibit at the show, obtained an AM. One of the complaints was of discrimination against the amateur; plus ça change . . .

C. 'Warley White' (which name we propose as the correct one, by reason of the Rules of Nomenclature, as well as the fact that it was awarded an AM under this name!) is after the style of a lax *C. carpatica*, with stems of some 20 cm which tend to tumble prostrately around the basal tuft. The sparse leaves are

53 *Campanula* 'Warley White'

heart-shaped like those of this parent, but are of a characteristic yellow shade. The flowers are semi- or fully double, and are about 4 cm across, open almost flat and held to the sun.

This plant has had a reputation for lack of hardiness, but this is far from the case, provided that it is so placed that winter dormancy is assured. This is best secured by a shaded or even northerly position, where the odd day of winter sun will not stimulate growth which will subsequently be scorched by frost. Any repetition of this rapidly exhausts a plant, and this has been recorded also for other small campanulas, such as *C.* 'G. F. Wilson' and *C.* 'Pseudoraineri'.

As mentioned, *C.* 'Warley White' obtained an AM when shown by Prichards of Highcliffe, Hants in 1925.

C. 'Warley' obtained a similar award in 1899; that time, actually shown by Miss Willmott! This was blue, and also a seedling of *C.* 'Haylodgensis', but we have not been able to trace its existence today.

C. × WOCKEI 'Puck' 10 cm Pale blue July

The parentage of this hybrid is probably *C. pulla* × *C. waldsteiniana*, although reliable accounts of its origin are hard to find. It is apparently closer to the former, and it is not impossible that it be a natural hybrid; for a brief discussion of this possibility, we refer to the entry on *C.* × *stansfieldii*, to which it is related.

This little plant, though not commonly offered, is still available today, and may even be found in more discerning Garden Centres.

A winter resting-bud of tiny wedge-shaped leaves arising from underground runners enlarges in spring to form a rosette of oval leaves which throws up a stem some 10 cm tall. This stem bears lance-shaped stalkless leaves, recalling those of the *C. waldsteiniana* parent, and terminates in a dangling pale blue bell.

The plant forms a slowly spreading diffuse mat, is reliably perennial, and is easily cultivated in an open soil with grit and some humus. It is also suitable for pot cultivation, but, like so many other campanulas, will be happier under these circumstances when repotted regularly each year.

This campanula seems always to be referred to as *C.* × *wockei* 'Puck', but it is difficult to trace the name 'Puck', in view of the fact that there are no records of any other *C.* × *wockei*.

C. ZOYSII Wulfen 8 cm Blue July/Sept

Although this does not count among the easiest alpines, it is so different that we have fallen into temptation, and so include it here. We cannot do better than quote Farrer: 'The last and strangest of the race—that minute exquisite rock-jewel which you may see filling the crevices and chinks of the Karawanken . . . rosettes of tiny spoon-shaped foliage, glossy and bright green . . . shoots of several inches carrying a number of long pale-blue bells so oddly bulging and puckered at the mouth as to resemble nothing on earth so much as a tiny soda-water bottle with a ham-frill at the end.'[2] Clarence Elliott suggested that the only answer to the slugs which find it so tempting is to grow it in such quantities that even the most gargantuan of slugs came all over bilious at the size of the beds of it.

It obtained an AM in 1924. This is really one for the specialist alpine grower.

54 *Campanula zoysii*

References

1. R. Farrer, *The English Rock Garden* (1918), p. 199.
2. Ibid., pp. 205–6.

Plant Associations

We must apologise for using fashionable jargon for the title of this chapter. But as a taxonomist friend immediately pointed out, the phrase 'plant association' does have a technical meaning. The hackneyed words came to life when he asked if we were 'writing a chapter about the plant communities which campanulas inhabit in the wild'. Well no, we're not. But what good advice that is for the gardener who wants to make attractive plant groupings.

In Beth Chatto's garden, famous for its harmonious and dramatic plantings, there is a drift of simple *Campanula persicifolia* in blue and white against a backdrop of great yellow gentian. A combination straight from an Alpine meadow. It is worth remembering that campanulas are in the main meadow or mountain plants, and have gentle colour tones.

In the herbaceous border

The classic herbaceous border is perhaps the epitome of English gardening. How many campanulas could be included in a list of plants of the highest quality for a summer border? They flower at the right time, but only two species are really the tops, with two others worth honourable mention.

First, *Campanula lactiflora*, tall and billowing at the back of the border, in pink, white or milky-blue, gains Graham Stuart Thomas's accolade as one of the 'finest of hardy perennials'. Yes, it may need careful staking, but it repays with grace, a delicate foam of colour, and long flowering. It will repeat flower, if cut back, and there is even a dwarf form in *C. lactiflora* 'Pouffe' for the foreground. *C. lactiflora* 'Loddon Anna' the pale dusty pink form, flowers at about 1½ m (5 ft). The equally unusual dark maroon-flowered scabious, *Knautia macedonica*, is a good foil at about 60 cm (2 ft). In the same colour range a drift of pink lavender will give solidarity to the waving flowering stems of the campanula and scabious.

Secondly, *Campanula latiloba* is a neglected but top rank plant for a classic border. It combines solid colour, sturdy stems, long flowering and a vigorous nature. The type plant is lilac-blue; in the cultivars *C. l.* 'Percy Piper' and *C. l.* 'Highcliffe' it is a rich, rather unsubtle lilac-purple. More attractive, and perhaps less invasive is *C. l.* 'Hidcote Amethyst', a strange night-time shade, deep mauve at the midrib of the petals, and paler and pinker at the edges. It is still grown in the central borders of the Old Garden at Hidcote Manor in a huge drift backed by spring-flowering shrubs and *Geranium magnificum*. The effect is sensational.

C. latiloba 'Alba' is grown nearby in a shady part of the white garden. It is brilliant white in colour, covered in flowers, strong and self-supporting. A plant with all the virtues that is surprisingly rare. At Hidcote it has *Rosa* 'Grüss an Aachen' and *Artemisia* 'Silver Queen' for company.

These two species then, would warrant a place in any herbaceous border. Two others are almost as good.

Campanula persicifolia at its best is a delicious plant. One of the strongest cultivars is *C. p.* 'Telham Beauty'. It is tall, large-flowered, and will repeat if cut down by half after the first flowering. It looks good with the more solid flowers and foliage of herbaceous peonies. It is a perfect complement to the rather awkward colour of the thornless *Rosa* 'Zéphirine Drouhin'. The bluey-pink of the rose finds a counterpoint in the mauvy-blue of the campanula. The more fancy forms of *Campanula persicifolia* just fail to make the grade. They have thin stems which are difficult to stake, and a capricious nature. They usually start easily enough, but they rapidly exhaust the soil and can fade away or fail to flower.

A much more reliable plant, often used in the background of an herbaceous border, is *Campanula latifolia*, in pale blue, purple or white. The strong purple of *C. l.* 'Brantwood' is a good contrast for *Aconitum carneum* which grows to about 1 m (3 ft) and is of pale old rose colour. *Campanula latifolia*, with its dramatic 1½ m (5 ft) spikes of large bells, flowers for about two weeks. This is the problem. It is rather a short time for a plant which will leave a large gap in the border, so it is best planted at the back, and in shade.

In the front of the border

If you take a robust attitude to plant snobs; if you want a simple colourful and easy plant with no staking and no fuss; then buy a packet of seed of *Campanula carpatica*. A heady mixture of blue and maybe white bells in saucer, cup or salver shapes will spring up. They make an effective edging to a border devoted to spring bulbs, taking over when the bulbs are finished. They look distinguished when mixed into a group of silver shrubs. They will take the front line of a rose bed instead of the traditional violas.

All of these uses and more apply equally to *Campanula portenschlagiana*. Its delicate deep blue bells are distinguished enough for a connoisseur's rockery. It is one of Will Ingwersen's favourites, and can only be said to flower too much. It is easy enough to be grown as a blue necklace around the white trunk of a specimen silver birch. In the terrace garden at Hidcote Manor it spills down a bank with brilliant orange and copper helianthemums. If you have a path leading straight from the gate to the front door it will make a ribbon of blue along its edge. Its brilliance is equally effective whether the house is a thatched cottage or a modern box.

These are the plants that pay rent. But many gardeners like a challenge and are prepared to search out something unusual. There are campanulas which may not make your visitors stop and gasp, but will have the more discriminating amongst them coo and beg a cutting.

Campanula punctata 'Rubrifolia' needs a little staking and a good soil. In its best form it is a strong maroony-red. It looks terrific near silver artemisias and the large furry leaves of *Salvia argentina*. The bells are large, there are many of them, and although it stands at about 45 cm (18 in) it is best at the front of the border where a visitor can be persuaded to lift a bell and admire the spots inside. The type plant has white bells which also have dark pink spots inside. Close up this is enchanting: but the effect from a distance is of a rather muddy white caused by the internal spots.

The hairy greyish foliage of *Campanula sarmatica* is especially good against stone. It is vigorous and easy enough for the front of a border. One with an edging of mowing stones, or a tiny wall would be best, as it arches forward gracefully, and is spoilt by staking. The new smallish blue hostas are a good foil, and like the same cool conditions. It is equally good grown like an alpine in a large rockery or stone wall. *Campanula versicolor* is an alpine plant from the mountains of Greece, but it grows to about 60 cm (2 ft) and is quite easy. It also looks right alongside stone, and grows from a single woody base. It can be used safely in a wall or rockery where the occasional large plant is needed to counter the dinky toy effect. Indeed Gertrude Jekyll even recommended this treatment for the 2 m (6 ft) *Campanula pyramidalis*: 'Its best place is a joint in a cool wall where it becomes perennial, and will probably seed itself. It is best to sow the seed in a limy compost in a joint low down and await the result.'[1]

In passing perhaps we should mention the traditional use of *Campanula pyramidalis*. Victorian gardeners grew the plant in pots, first in a cold frame, and finally after potting on several times, in the vinery or conservatory. After about 18 months the plants flowered and could be 2 m (6 ft) in height. The pot was then carried into the house and used to decorate the empty chimney piece of hall or drawing room in the summer. Louis Liger wrote in 1706 'place it upon the half space before a chimney, with a Pot of Tuberose on each side of it and a Pot of Scarlet Lychnis on each side of the Tuberose' and recommended 'small pots of Sweet Basil and Origano neatly ranged in front'.[2]

The paint pot

To recommend colour is to step into a minefield: one gardener's magenta is another's rosy-pink. But it ought to be said that the campanula blue is best described as Spode—it has a bias towards lilac. If it meets the pure greeny-based blue of, for instance, a delphinium it is a hit below the belt.

Campanulas *latifolia* and *latiloba* are used extensively in the famous double blue and yellow borders of Clare College, Cambridge. Here herbaceous plants are grown to perfection. The discerning use of just two colours (and green) has produced a result which is surprisingly exuberant. The campanulas associate delightfully with *Alchemilla mollis* (ladies mantle), *Galega officinalis* (goats rue), *Lysimachia punctata* (yellow loosestrife) and *Tradescantia* (spider wort). In most years the mauve-blue of *Campanula latifolia* is finished before the delphiniums start. If they do overlap it is not a success.

The successful blue/yellow scheme is possible on a smaller scale. *Campanula glomerata* 'Superba' can be underplanted with golden lamium (*Lamium maculatum* 'Aureum'). In shade at Hidcote Manor, *C. latiloba* has a background of yellow oxalis.

The other colour to watch out for with campanulas is the scarlet of modern roses. Campanulas are much more successful with the pale pinks and dark bluey-reds of old fashioned roses.

White is also a colour. It has been very fashionable since Lawrence Johnson and Vita Sackville-West made their white gardens. This is the place to show off some of the double white forms of *Campanula persicifolia. C. p.* 'Alba Coronata', *C. p.* 'Hampstead White' and *C. p.* 'Boule de Neige' are all distinctly different. They are vigorous enough to make a good show provided they are split and replanted annually. Dark green foliage, such as that of *Malva moscata* 'Alba', or the tall spike leaves of *Iris orientalis*, provide a good foil. There are never enough large leaves to counter the bitty effect of small herbaceous flowers. *Crambe maritima* (sea kale) with purplish-green leaves, or *Silybum marianum*, whose leaves are dramatically splashed white and green, form a cool background for the flowers. So will dark green or blue hostas, of course.

The pure shining whiteness of *C. latiloba* is indispensable for a white border. It has the added bonus of dark evergreen foliage. *Campanula latifolia* 'Alba' will glow in the dusk, if your white border has a shady side. The rare *Campanula trachelium* 'Alba Flore Pleno' has showy flowers rather like fussy lampshades. It will form a good sized clump about 60 cm (2 ft) high and wide given the opportunity. But it does not often get a season undisturbed as visitors often beg a piece, and the plant has a woody rootstock which is a beast to propagate. The grey-felted-leaved shrub *Phlomis* 'Lloyd's Variety' which does not flower, provides a comfortable rounded mass against which to set this rather frilly campanula. The larger and darker leaves of *Phlomis* 'Edward Bowles' would do the same job: it might be necessary to remove the yellow flowers. White siberian irises would be a pleasant addition to the scheme.

In the woodland

In *The English Flower Garden*, William Robinson bluntly recommends 'the tall and straggling kinds [of campanula] admirable for the wild garden, or rough woody place or hedgerows, but must not be used much in the Flower Border as their time of bloom is short . . . and they are apt to overrun rarer plants.'[3] This good advice applies especially to *Campanula latifolia, C. trachelium* and *C. alliarifola*. They seed themselves mildly. *Campanula trachelium* looks fine on an ivy-clad bank, or under a hedge. At Wisley it is allowed to seed itself about in amongst rhododendrons and primulas in the woodland behind the rock garden. Gertrude Jekyll specified *C. alliarifolia* for a woodland path, or in that no-man's land between the grass and open woodland in a large garden. It is quite a graceful flowering plant with good leaves, but just fails to be first class since it does not open its white foxglove-like flowers wide enough.

Chance often has a role to play in gardening. Visitors to Hidcote Manor in Gloucestershire a few years ago will remember the Spring Slope. In open woodland hundreds and hundreds of blue and white *Campanula latifolia* had seeded themselves. They covered perhaps 2,000 sq m ($\frac{1}{2}$ acre) under the canopy of tall trees. For two weeks they were the queen of the garden. Recent reconstruction has disturbed them, but a few remain which may furnish another colony. Perhaps an alkaline soil is needed to enable *C. latifolia* to flourish under these conditions, for Beth Chatto, in Essex, has used the white form in a similar way to lighten the back of a large border in deep shade under oaks. It shines like a good deed in a naughty world.

The sixteenth century

It is possible to be spoilt for choice, and it is difficult for a keen plant lover to say no to an interesting novelty. Restricting the choice by date may be an interesting exercise. In some gardens this is done by only using those plants which were available up to the date that their house was built and during its heyday. The Marchioness of Salisbury has done this with the sunk garden at Hatfield House in Hertfordshire: the formal gardens around the house all contain old fashioned plants. The sixteenth century gives us *Campanula medium* (Canterbury Bells), *C. trachelium, C. persicifolia* and *C. pyramidalis.* All of these look good with herbs like rosemary, rue and lavender. The big full-petalled flowers and large leaves of *Paeonia officinalis* are an attractive background to them. Gerard mentions the double forms of *Campanula trachelium*. Plant then with the peerless white Madonna Lily, *Lilium candidum*. Give them the very best start you can, and then leave them undisturbed.

Some of our roses predate even these early campanulas. *Rosa alba* goes back to Roman times, and so does *Rosa* 'Autumn Damask. *Rosa mundi* is supposed to be associated with Fair Rosamund of Henry II. If the date is less important than an authentic Elizabethan atmosphere, then many of the old shrub roses will complement *Campanula persicifolia* or *C. medium*. In this case the campanulas should be restricted to nature's single forms in white or blue.

An eye for a plant

We leave the last words on the subject to E. A. Bowles, writing in *My Garden in Summer*. He loved his plants, and had the leisure to study and observe them. Here he describes *C. lactiflora*, the tallest campanula which he allowed to take over part of the rock-garden.

'The greatest change that I always notice in the rock garden on my return from the hills is the final two feet of growth and bursting into flower of *Campanula lactiflora*, which has become one of our weeds there, and if it were not ruthlessly evicted wherever it is not required would cover the whole place, including the moraines . . . I planted three seedlings, all I had, and if any clairvoyant had been crystal-gazing at the moment and told me I should in a few years be digging them up in hundreds to give away and throw away, I should have dug up those three infants, fearing the ruin of my garden. But as I destroy those which . . . stand where they should not, only leaving a row down each side of the path and a few others where they are doing no harm, the rock garden looks exceedingly well during the fortnight of their reign. Nearly all of them at this upper end of the rock garden are either the pure white form, which I think the loveliest of all, or of the skim-milk, bluish-grey tint that provided their specific name, and really blue forms very seldom appear among them. In some seasons the five or six feet high flower-stems they produce have caused trouble where they grow beside the path by arching over, after a thunderstorm, until they meet in the centre, and the first who ventures to walk through them gets a shower-bath. So now I prepare for such emergencies by nipping off the heads of the stems nearest the path when they are only two or three feet high, which causes them to branch freely, and yet they flower at the same time as their untouched brethren behind, but on shorter stems that do not bend out so far, and have a very pleasing effect in the forefront. It is marvellous how soon *C. lactiflora* seems to change from lovely flower-heads eighteen inches high by ten through into clusters of pepper-pots shaking out thousands of minute flattened buff seeds with every jolt and jar . . . So as soon as the majority of the flowers of a head look a little jaded it is wise to cut off the whole head, or you must prepare for a year or two's extra weeding of seedlings. If cut off close under the lowest flowers the stems will branch out and flower again later in the season; but it is a poor show that is provided by these small heads of lateral shoots compared with the waving masses of early July.'[4]

References

1. G. Jekyll, *The Garden* (May 1922), p. 210.
2. A. M. Coats, *Flowers and Their Histories* (1956), p. 41.
3. W. Robinson, *The English Flower Garden* (8th edn, 1901), p. 462.
4. E. A. Bowles, *My Garden in Summer* (1914), pp. 208, 209.

Campanulas for Cutting

In country gardens there have always been flowers grown for cutting. Step over the threshold of a large country house at the turn of the century: naturally there would be campanulas in the house in June and July; perhaps an enormous potted chimney bellflower (*C. pyramidalis*) in the hall. The flower vases would be quite different from those of the modern flower arranger. The head gardener would have grown his campanulas to perfection, and only brought the finest spikes into the house. He would often arrange them himself, probably in straight-sided vases, and only one variety in each vase. Mixed flowers were considered vulgar. The flowers were grown to display the gardener's skill, and treated quite plainly as if at a show. In fact, there was a great deal of rivalry between neighbouring establishments as to whose gardener produced the earliest peaches, largest grapes and the finest flowers.

C. persicifolia is the only real 'florist's flower' amongst campanulas which would be worthy of a vase to itself.

It has been highly bred in the past to produce exotic varieties. It was also grown in very large quantities for the market before the First World War, and in the 1920s. Both single and double varieties make excellent cut flowers; they are approximately 60 cm (2 ft) high. Ideally, the plants should be grown in well spaced pairs of rows, 45 cm (18 in) apart with about 80 cm (30 in) between the pairs of rows. The double row helps the plants to stand up well, though even so they may need staking. The larger space on either side gives easy access. The plants should be lifted and split every other year. Good loamy moist soil such as in the vegetable garden will produce the best stems. This cultivation would suit any of the campanulas with mildly running roots. The flowers of *C. persicifolia* are fragile and easily marked by bad weather, but they will travel if carefully packed. A watch should be kept for rust. Like most campanulas they last quite well in water. They are best picked in the cool of the morning or evening. The milky sap, inulin, which the cut stems exude can be a nuisance. It helps to singe the stems. They should then be plunged in cold water for a long drink before using.

Sad to say, those days of flowers grown in walled gardens for the big house have gone forever. There are not even many households today which can allocate an area of the vegetable garden to flowers for the house.

Gardeners who want to provide flowers for the house, will usually do so from their borders. Indeed the herbaceous border will produce magnificent blooms if the owner can bear to cut them. It is still possible to see bunches of mixed border flowers on the market stalls of country towns. They are a delightful higgledy-piggledy mixture of many different varieties of old world flowers, and are quickly snapped up.

In June and July quite a range of campanulas will appear in these mixed bunches. The easiest is *Campanula glomerata*, in its strong growing forms of *C. g.* 'Superba' or *C. g.* var. *dahurica*. The stems are straight, and with their flowers in a cluster at the top they are easy to bunch and to arrange. They should be cut when only half open, when they will last very well. The rather shorter-lived *C. latifolia* and *C. lactiflora* will make quite an impact in a large arrangement: but these will droop quickly if kept out of water. Perhaps one of the best is *C. latiloba*, which has enough impact to look good on its own, or in a mixed bunch. It lasts quite as long as its relative *C. persicifolia* but has the advantage of stronger stems, and many flowers to each spike. *C. trachelium* is another easy plant that is good for an informal mixture, though it is rather too coarse to have a vase to itself. The two double forms of *C. trachelium* in lilac-mauve and white are much finer. All these last should be plunged up to their necks in water overnight, or they may not take up water very well.

If *Campanula rapunculoides* is making a bid to take over the garden, revenge can be taken by cutting it. This will have the added advantage of preventing it from seeding. Its spikes of quite large flowers are very graceful.

Large informal mixed arrangements are still with us in country houses. It little matters that the stems are not quite straight, nor the colours co-ordinated—the charm is in the diversity and profusion of flowers.

Modern flower arranging is rather different. Members of NAFAS (National Association of Flower Arranging Societies) are interested in the form and shape of plant material as well as the flowers. This has had a significant effect on the conservation of unusual rare plants. The floral designs are often minimalist, reflecting an interest in Eastern Arts, and also perhaps today's smaller gardens, and the expense of florists' flowers. The large herbaceous borders have mostly gone. In their place are mixed borders, where the enthusiastic plantsperson will be growing specialist plants. There are campanulas which will provide the 'something different' which is exactly what the flower arranger craves.

Cream-white *C. alliariifolia* immediately springs to mind as a suitable subject for an Ikebana arrangement. The stems are held gracefully on the plant, often splaying out and then turning up at the tip. They form the curved stem which can provide an interesting angle to juxtapose with twigs or foliage. The leaves are a decorative grey-green heart-shape, and softly hairy. They are rather large, and a few would undoubtedly need to be removed. Other plus points are *C. alliariifolia*'s ease of cultivation, its repeat flowering, and the way it produces a few self-sown seedlings to perpetuate itself.

It is a pity that *C.* 'Burghaltii' is not as easy. This has all the grace of stem form that *C. alliariifolia* can provide, but is more distinguished. The larger, narrowly drooping bells are a most curious amethyst-grey. It would be a pity to mix this distinctive plant with other flowers. The problem would be to find enough cutting material, though it does repeat flower for quite a long period.

Along the same lines *C. punctata* is rather easier to cultivate and would produce more stems per plant. The bells are about the same large size as *C.* 'Burghaltii' and also hang gracefully. They are white, wonderfully spotted with red inside, and in a well-placed arrangement their full beauty would be apparent: it is not often easy in the garden to look into the narrow tubular bells. Both these plants take water quite well, and benefit from an overnight drink before being handled.

The new *C. takesimana* is rather similar to the white form of *C. punctata*. However, it runs quite a bit and would certainly make plenty of delicate graceful stems, with their large red and white flowers. This plant's fragile appearance belies its tough nature, and can stand quite a bit of hacking about. All these three need sensitive staking in the garden or they will flop. In a Japanese-style arrangement their sprawling curving stems would be an advantage.

C. carpatica is a plant with one foot in the border and the other in the rock-garden. Quite short, about 30 cm (1 ft), its flowers are large for the length of stem. This is often the way with alpines, and it makes them difficult to use in a vase. But *C. carpatica* makes up for it by producing flowers by the hundred, on fairly strong wiry stems. Cutting the flowers is a positive advantage for it ensures that it will repeat flower in the autumn. A single plant will provide dozens of simple, innocent cup-shaped bells; their removal would go almost unnoticed. *C. carpatica* would be a boon for table decorations at a summer wedding. The ubiquitous *C. portenschlagiana* and *C. poscharskyana* are equally giving plants: their flowers are smaller, starry and held in sprays.

This brings us to the miniatures. Many of the alpine campanulas have tiny flowers which would be the making of one of those little arrangements which always seem to be a feature of flower arranging competitions. However, they can be a challenge to grow, and could be killed by injudicious cutting. The following are a few which combine ease of cultivation, with the delicacy of scale that is required.

C. patula is a very worthwhile biennial. Keeping it going by self-sown seedlings is not a problem. The result is a harebell, which is so profuse in its flowering stems, that no flower arranger could keep up with it. The stems are strongly wiry, the flowers starry bells. *C. barbata* is very distinctive. The flowers are a fine China blue in a good form, intensely hairy inside and out and have a yodelling alpine look. The plants are short-lived, but a strong specimen can produce a lot of stems. We found they lasted well in water, when some wind-damaged stalks were brought into the house.

Lastly, two real miniatures which are just as easy as their border counterparts. *C. cochleariifolia*, Fairies Thimbles, runs about mildly and is profusely flowered, but only 10 cm (4 in) high. Its double form *C. c.* 'Elizabeth Oliver' is minutely fluffy, and a pale Spode blue. Deep Tyrian purple is a visual surprise when it comes on a flower not 1 cm ($\frac{1}{2}$ in) long. *C. pulla* is a runner which potters around other plants, often coming up away from the original position. For all that, it is prolific with its flowers, and should give the maker of a miniature scheme a strong colour to conjure with.

Appendix I '*The Plant Finder*'

The Plant Finder, published annually by the Hardy Plant Society, gives lists of campanulas and where they may be obtained. As our book is also sponsored by the Society, we expect that Members may wish to cross-refer to it, and we have therefore tried to cover most of the plants mentioned in the current edition. The following short lists are of plants which we have not included in this volume, with the reasons why.

Specialist Alpines, outside the scope of this book:

C. alpina	*C. hawkinsiana*	*C. orphanidea*
cenisia	*hercegovina*	*petraea*
choruhensis	*lasiocarpa*	*piperi*
coriacea	*lyrata*	*shetleri*
crispa	*morettiana*	*thessala*
davisii	*oreadum*	*troegerae*

Names probably incorrect, doubtfully valid or without adequate description

C. 'Elizabeth Orange'
kladniana = *rotundifolia* group
kolenatiana
'Yvonne'

Appendix II *RHS Awards*

Over the years, the genus *Campanula* has received the following awards from the Royal Horticultural Society. Unless otherwise stated these are Awards of Merit (AM); AGM = Award of Garden Merit; FCC = First Class Certificate; PC = Preliminary Commendation.

Name of plant	*Date*	*Shown by*
C. abietina	1891	Paul
'Abundance'	1915	Grove
acutangula	1915	Prichard = *arvatica*
allionii 'Alba'	1930	Jenkin = *alpestris alba*
allionii 'Frank Barker'	1930	Elliott = *alpestris* 'Frank Barker'
alsinoides	1932	May
alpestris	1984	RBG, Kew
andrewsii	1934	Hanbury
argyrotricha	1932	May
arvatica	1952	Peacock
arvatica 'Alba'	1937	Elliott
atlantis	1952	Ingwersen
aucheri	1960	Buchanan
aucheri 'Quarry Wood'	1965	Simmons
'Balchiniana'	1896	Balchin
barbata	1951	Hammer
betulifolia	1937	Boothman
'Birch Hybrid'	1945	Ingwersen
broussonetiana	1957	RBG, Kew = *lusitanica* (annual)
'Brookside'	1933	Brookside Nursery
caespitosa	1951	Ingwersen = *cespitosa*
calcicola	1923	Bulley
calaminthifolia	1936	RHS Gardens
carpatha	1952	Saunders
carpatica 'Blue Bonnet'	1968	Gault
'Bressingham White'	1967	Bloom/Gault

Name of plant	Date	Shown by
C. carpatica contd 'Chewton Joy'	1929	Prichard
'Giant'	1931	Prichard
'Grandiflora'	1967	Aslet
'Harmony'	1931	Prichard
'Loddon Fairy'	1967	Gault
'Mrs V. Frare'	1930	Prichard
'Queen of Somerville'	193?	Prichard
'Slaugham White'	1937	Blundell
'The Pearl'	1967	Gault
'White Convexity'	1967	Gault
'White Star'	1905	Prichard
cashmeriana	1958	Lamming
cenisia 'Alba'	1914	Tucker
'Chastity'	1916	Grove
'Clarence Elliott'	1937	Elliott
cochleariifolia 'Miranda'	1920	Eliott and AGM 1935
crenulata	1926	Bulley
'Donald Thurston'	1926	Thurston
elatines	1933	Giuseppi
elatinoides	1949	Saunders
'Enchantress'	1918	Grove
ephesia	1956	Ingwersen/RHS
excisa	1933	Giuseppi
'Fergusonii'	1904	Ferguson
formanekiana	1931	Giuseppi
fragilis (FCC)	?	Tjaden
garganica (AGM)	1930	?
fenestrellata	1950	Griffiths
'W. H. Payne'	1914	Watson
glomerata dahurica	1965	Gault
'Superba'	1954	Ingram
grandis 'Highcliffe'	1935	Prichard = *latiloba*
'Gremlin'	1946	RHS, Wisley
'Hallii'	1923	Hall
hawkinsiana	1932	Bevan
'Hendersonii' (FCC)	1885	Ware
hercegovina	1933	Lawrence
'Nana'	1946	Ingwersen
incurva	1937	RBG, Kew
'Isabel'	1904	Prichard
isophylla 'Alba' (FCC)	1888	Ware
'Mayi'	1899	May

Name of plant	*Date*	*Shown by*
C. kolenatiana	1918	Tucker
laciniata	1945	Ingwersen
lactiflora (AGM)	1926	
'Coerulea'	1901	Perry
'Loddon Anna'	1952	Carlile
'Pouffe'	1966	Bloom
'Prichard's Variety'	1964	Bloom
'Superba'	1969	Gault
lasiocarpa	1947	Weir
latiloba (AGM)	1936	
longestyla	1967	Lawrence
'Lynchmere'	1948	Panckridge
medium 'Calycanthema'	1889	Veitch
'Flore Pleno'	1889	Veitch
'Flore Pleno Roucarmine'	1929	Haage & Schmidt
'Mauve'	1929	Hurst
'Meteor'	1915	Grove
'Single White Improved'	1971	Clause
'White'	1929	Webb & Dobbie
mirabilis (FCC)	1898	
mollis	1932	Brammall
morettiana	1932	Elliott
'Alba'	1934	Stoker
'Eximia'	1934	Stoker
nitida 'Alba'	1970	Finnis = *persicifolia* 'Planiflora Alba'
nobilis × 'Norman Grove'	1914	Grove
oreadum	1940	Crook
'Pamela'	1953	Saunders
pelia	1950	Crook = *thessala*
'Percy Piper'	1965	Bloom
peregrina	1903	Veitch
persicifolia 'Alba Grandiflora'	1890	Paul
'Coronata Shirley'	1925	Ladhams
'Fleur de Neige'	1921	Miller
'Frances'	1938	Holding
'Moerheimii'	1900	Ware
'Ryburgh Bells'	1923	Stark
'Spetchley'	1921	Berkley
'Telham Beauty'	1916	Barr and AGM 1928
'The King'	1923	Ladhams
'Wedgwood'	1953	Watkins & Simpson

Name of plant	*Date*	*Shown by*
C. persicifolia 'William Lawrensen'	1907	Lawrensen
petrophila	1952	Lilley
phyctidocalyx	1909	Lawrence
'Phyllis Elliott'	1918	Elliott
pilosa 'Dasyantha'	1950	Heath = *chamissonis*
portenschlagiana (AGM)	1927	
poscharskyana	1933	Giuseppi
'Profusion'	1896	Jenkins
propinqua 'Grandiflora'	1931	Cecil
pulla	1976	Earle
pusilla 'Miss Willmott'	1915	Eliott = *cochleariifolia* 'Miss Willmott'
pyramidalis 'Alba'	1896	Wythes
'Compacta'	1892	Wythes
raddeana	1908	Reuthe
'R. B. Loder'	1922	Prichard
raineri	1970	Earle
'Rotarvatica'	1935	Tyler
rotundifolia 'Jenkinsii'	1922	Jones
'Olympica'	1931	Watkins & Simpson
rupestris	1904	Cutbush
(unnamed form)	1967	Barton
rupicola 'Giuseppi's variety'	1937	Giuseppi
saxatilis	1933	Giuseppi
scheuchzeri 'Covadonga'	1939	Elliott = *carnica* 'Covadonga'
spathulata 'Giuseppi'	1933	Giuseppi
stevenii 'Nana'	1913	Prichard
tomentosa 'Maude Landale'	1914	Landale = *rupestris*
trachelium 'Bernice' (PC)	1965	Gault & Bloom
tridentata	1935	Baker
tubulosa	1933	Giuseppi
versicolor	1932	Baker
vidalii	1960	RBG, Kew = *Azorina vidalii*
'Warley'	1899	Willmott
'Warley White'	1925	Prichard
'Woodstock'	1920	Jenkins
zoysii	1924	Tucker

It is a matter of regret that so many of these are rare in cultivation today, and even more so that some are unknown and unobtainable.

While it's not easy to make these observations in the short time you often have to watch a "mystery bird," practicing these methods of identification will greatly expand your skills in birding. Also, seek the guidance of a more experienced birder who will help you improve your skills and answer questions on the spot.

Bird Songs and Calls

Another part of bird identification involves using your ears. A song or call can be enough to positively identify a bird without seeing it. If you see a bird that you don't know, the song or call can help you identify it. To learn about bird songs, calls, how they are produced, what they mean and how they are used, see the companion *Birds of New York Audio CDs*.

Appendix III *Societies and Sources of Seed*

The annual seed lists of all the following usually contain the names of campanulas of all sizes. The seed is collected by members (with the exception of the RHS) and is then circulated to other members on request free of charge, except for postage.

Alpine Garden Club of B.C.
Box 5161—MPO, Vancouver, B.C. V6B 4B2, Canada

Alpine Garden Society
The Secretary, Lye End Link, St Johns, Woking, Surrey GU21 1SW

Cottage Garden Society
Membership Secretary: Mrs Philippa Carr, 15 Faenol Avenue, Abergele, Clwyd LL22 7HT

Hardy Plant Society
Membership Secretary: Mr S. M. Wills, The Manor House, Walton-in-Gordano, Clevedon, Avon BS21 7AN

Royal Horticultural Society, 80 Vincent Square, London SW1P 2PE

Royal Horticultural Society's Gardens, Wisley, Woking, Surrey GU23 6QB

Scottish Rock Garden Club
Honorary Secretary: K. M. Gibb, 21 Merchiston Park, Edinburgh EH10 4PW

SOCIETIES IN THE US

Hardy Plant Society
124 S.W. Meade, Portland, Oregon 97201

American Rock Garden Society
Secretary, Buffy Parker, 15 Fairmead Road, Darien, CT 06820

Glossary

ALTERNATE leaves attaching to the stem singly (not in pairs).
AMPLEXICAUL the manner in which a sessile leaf clasps the stem; uncommon in *Campanula*.
ANTHER pollen-bearing organ of flower.
ANTHESIS time of flowering.
APPENDAGE growths in the gaps between the calyx lobes.
AXIL the angle between a branch or leaf and the stem which give rise to them; generally the upper one, in which a bud often originates.
AXIS (imaginary) line running through the centre of an organ.
BASAL the leaves at soil level.
BRACT a much-reduced leaf at the base of a flower.
CALCIPHILE lime-loving: or at least growing on limestone.
CALCIPHOBE lime-hating; growing on acid soils.
CAPSULE a dry fruit, or seed-case, which splits to shed seed.
CALYX outer whorl of flower; in *Campanula*, made up of five sepals or calyx lobes (and with or without appendages, according to species).
CAMPANULATE bell shaped; in our genus, opposed to infundibular, rotate or tubular, which see.
CARPEL stigma, style and ovary as a total structure.
CAULINE pertaining to, or borne on, the stem and above ground level, generally referring to leaves.
CLONE a plant produced from another single plant by vegetative propagation; members of a clone all have identical genetic make-up.
CORDATE heart-shaped; correctly referring to the leaf base, but frequently to the whole leaf.
COROLLA the whole whorl of petals.
CRENATE rounded marginal teeth of leaf.
CRENULATE diminutive of previous.
CYMOSE inflorescence or flower-cluster with each side stem ending in a single flower, opening successively from the tip.
DECIDUOUS dropping off at end of season; as opposed to evergreen.
DECUMBENT prostrate (of stems) but with ascending tips.
DEHISCENCE splitting open when mature (anthers, capsule, etc.).
DENTATE teeth of leaf-margin directed radially outward.

DENTICULATE diminutive of foregoing.
DISTAL remote from the attachment, as opposed to proximal.
ENDEMIC growing only in a given area.
ENTIRE (leaf) without teeth at margin.
ERECT upright carriage (usually of capsule); opposed to pendent.
EXSERTED (style) projecting beyond tips of corolla.
FLEXUOUS (leaf or stem) wavy.
GLABROUS smooth; without hairs, spines, etc.
HERBACEOUS not woody.
INCLUDED not projecting; shorter than corolla.
INFLORESCENCE whole flower-cluster.
INFUNDIBULAR funnel-shaped, where the flower is not bell-shaped nor the petals reflexed.
LANCEOLATE lance-shaped.
LAMINA (leaf) blade.
LINEAR long, narrow, with parallel margins.
LOBE (corolla or calyx) the individual segment.
LYRATE a multi-lobed leaf with the rounded terminal lobe much larger than the side lobes.
MARGIN (leaf) edge.
MONOCARPIC flowering but once, dying subsequently.
MULE a somewhat vernacular term describing a non-fertile hybrid, just as a mule is the infertile progeny of a horse and a donkey.
OBLONG rectangular with rounded corners.
OBOVATE egg-shaped, with narrower end nearest stem, two-dimensionally.
ORBICULAR flat and circular in outline.
OVATE egg-shaped with wider part nearest stem, two-dimensionally.
OVOID egg-shaped, three-dimensionally.
OVULE seed before fertilisation.
PANICLE repeatedly branched inflorescence.
PEDICEL flower-stalk.
PERIANTH petals + sepals, collectively.
PETIOLE leaf-stalk.
PORES small openings (generally in leaf surface).
PROSTRATE lying flat on ground.
PROXIMAL nearest to point of attachment.
PUBESCENT hairy as a general term.
RACEME single flowers attached to the stem by a pedicel.
REFLEXED bent downward, usually abruptly.
RHIZOME prostrate underground stem, tending to root and throw up stems at the nodes.
ROSETTE cluster of leaves at ground-level.
ROTATE wheel-shaped; flat, circular corolla; in *Campanula* often star-shaped.

SCABROUS surface roughened by wartlike growths.
SEPAL lobe, or segment, of calyx.
SERRATE (leaf) having sharp forward-facing teeth.
SESSILE (leaf) without a footstalk.
SINUS cleft between lobes of petals or sepals.
SPIKE single flowers attached directly to the stem.
STAMEN pollen-producing organ of flower.
STELLATE star-shaped.
STIGMA part of style receptive to pollen.
STOLON slender modified stem which runs along the ground, often rooting at the nodes, as in the strawberry.
STYLE the long upgrowth from the carpel, the tips of which are the stigmas (usually in threes or fives in *Campanula*).
SUB- prefix meaning 'almost'.
TAXON any named taxonomic category (eg. genus, species, cultivar).
TRIFID (style) split into three parts.
TUBULAR in campanulas refers to the flower shape which is a long tube-shaped bell, often with flaring tips to the petals.
TYPE the originally described plant to which the given name has been attached.
TYPE LOCALITY the place from which a given plant was described.
UNDULATE wavy.
WHORL ring of organs.
WINGED usually refers to the leaf-stalk which is not round but is flanged along its length, the effect being of a downward thin continuation of the leaf surface. (As in the Primrose.)

Bibliography

Bailey, L. H., *A Garden of Bellflowers*, New York, 1953

Beddome, Col. R. H., 'An Annotated List of the Species of *Campanula*', *RHS Journal*, 1907

Bloom, A., *Hardy Perennials*, London, 1957

——, *Perennials for Trouble-free Gardening*, London, 1960

——, *Alpines for Trouble-free Gardening*

Boissier, E., *Flora Orientalis*, Vol. 3, Geneva, 1875

Bowles, E. A., *My Garden in Summer*, London, 1914 and revisions

Clapham, Tutin and Warburg, *Flora of the British Isles*, CUP, 1952

Clifford Crook, H., 'Campanulas: Their Cultivation and Classification', *Country Life*, 1951

——, *Campanulas and Bellflowers in Cultivation*, London, 1959

Correvon, H., 'The Genus *Campanula*', *The Garden*, 1901

Davis, P. H., *Flora of Turkey*, Vol. 6, Edinburgh, 1978

de Candolle, Alphonse, *Monographie des Campanulées*, Paris, 1830

Farrer, R., *The English Rock Garden*, 1918

Gray, Asa, *Synoptical Flora of North America*, Vol. II, Part 1, New York, 1878

Grey-Wilson, C., *Alpine Flowers of Britain and Europe*, London, 1979 and revisions

Hardy Plant Society, *The Plant Finder*, 1989

Harkness, M. and d'Angelo, D., *The Seedlist Handbook*, Timber Press, 1986

Hills, L. D., *The Propagation of Alpines*, 1959/1976

Huxley, A., *Mountain Flowers in Colour*, London, 1967 and revisions

Huxley, A. and Taylor, W., *Flowers of Greece and the Aegean*, London, 1977

Ingwersen, W., *Manual of Alpine Plants*, Eastbourne, 1978

Polunin, O., *Flowers of Europe, A Field Guide*, London, 1969

——, *Flowers of Greece and the Balkans*, Oxford, 1980

Prichard, M., 'The Genus *Campanula*', *RHS Journal*, 1902

Robinson, W., *English Flower Garden*, London, 1883 and revisions

Roemer and Schultz, *Systema Vegetabilium*, 1819

Shishkin, B. K., *Flora of the USSR*, Academy of Sciences of the USSR, translated from the Russian by the Israel Program for Scientific Translations, Jerusalem, 1972

Sibthorp and Smith, *Prodromus Florae Graecae*, 1806
Tutin, T. G. et al., *Flora Europaea*, Vol. 4 CUP, 1976
von Bieberstein, F. A. Marschall, *Flora Taurico-Caucasica*, 1808–1819

Index

Bold type indicates the main reference.